Unveiling the Fabric of Spacetime: Exploring Noncommutative Quantum Mechanics

Johnny

Contents

Chapter 1

Introduction

In recent years, there has been great interest in the study of deformations of Quantum Mechanics (QM) in order to understand how fundamental features are altered with these modifications. In particular, additional commutation relations on position and momentum operators have been imposed, either separately [5–7] or simultaneously [8–12]. This particular kind of deformation of QM stems from a result in string theory stating that the dynamics of the strings can be described by a gauge theory in noncommutative space [13]. However, interest in additional commutation relations arose much earlier, with the possibility of momentum noncommutativity being a good candidate for the regulation of divergences in quantum field theory since it would introduce a fundamental momentum scale, acting as a cut-off for integration on the momentum space [14]. In this work, the motivation is mainly attached to the former result as there is the expectation that a final theory of quantum gravity will unveil some noncommutative structure of the spacetime [15]. In this regard, the study of noncommutative Quantum Mechanics (NCQM) acts as a probe to the effects of such structure on low energy physics, by evaluating its effects on quantum mechanical systems and, additionally, providing for some hints as to where to look for evidence of these effects.

In this context, several works have delved into the consequences of this new structure, mainly on two distinct areas: one on the development of mathematical tools that are better suited for this particular task and the other as the analysis of the consequences of the NC structure on particular systems or phenomena. Concerning the former, the development of a phase-space formalism for NCQM (PSNCQM) in Ref. [15] based on its symplectic structure allowed for a sound mathematical base for the study of several phenomena, such as the Robertson-Schrödinger uncertainty principle and Osawa's uncertainty relation [16–18]. Furthermore, it allowed for a covariant formulation of the phase-space formulation of quantum mechanics. As for the consequences of the new commutation relations in particular systems, many features have been studied, in particular quantum entanglement and locality [19, 20], quantum decoherence [21–23], the gravitational quantum well for ultra-cold neutrons [11], gauge invariance [24, 25] and the equivalence principle [24]. Besides these systems, the implications of NCQM were studied in the framework of quantum cosmology, revealing that the singularities in black holes can be somewhat regulated in this context [26, 27]. Many of these results arise from theoretical analyses of the consequences that the NC parameters have on phenomena such as locality or entanglement and others are able to provide estimates on the scale of these parameters by comparison with experimental data. Hence, not only this deformation is a great tool for understanding how some QM systems behave upon deformations, but it also provides us with insights to the scale where new phenomena in physics might appear.

In light of these motivations, a continued and exhaustive study of the predictions of NCQM is in order. For that reason, this work considers some fundamental aspects of NCQM such as the no-cloning theorem and the process of teleportation [1] and the modification that arises in the relativistic dispersion relation due to the breaking of Lorentz invariance for NC space-time [4]. Additionally, in the spirit of Refs. [26, 27], we study the collapsing of null thin shells in the context of the canonical quantization of General Relativity (GR), i.e. Wheeler-deWitt (WdW) equation, when taking the NC algebra to the phase-space of metric components [3]. Within the same framework (WdW), we study the behaviour of a FRW-type metric in the context of quantum cosmology in Hořava-Lifshitz gravity, which is an ultraviolet complete extension of GR [2]. Here the effects of the quantization are studied by analysing the quantum to classical transition of this model.

For the purpose of introducing the required background for the following chapters of this work, the next sections are dedicated to the exposition of the necessary mathematics and physics frameworks that will be necessary. First, the phase-space formulation of quantum mechanics is introduced in detail. The equivalence with other quantization procedures is shown, as well as the prescription to quantize a classical system in this framework. Then, the noncommutative version of quantum mechanics is introduced, as well as its phase-space formulation. In order to do this, a brief review of symplectic geometry is given, together with its connection to classical and quantum mechanics. Besides the background in QM, some topics of GR are also necessary. Thus, the $3 + 1$ decompositon of a manifold is presented, and its use for the Hamiltonian formulation of GR. This leads to the introduction of the WdW equation. A brief study of the Schwarzschild black hole solution is also given. Finally, the Israel formalism for the patching of two solutions of the Einstein field equations is presented for the particular case of null hypersurfaces as the contact region. This is then applied to the collapse of thin null shells.

The ensuing chapters are then organized as follows. In Chapter 2, the no-cloning theorem in the context of PSNCQM is discussed, as well as the teleportation procedure and its fidelity for arbitrary states. Chapter 3 adresses the issue of Lorentz invariance violation by the noncommutative algebra and explores the Extended Phase-space formalism to derive a modified relativistic dispersion relation. Additionally, Gamma Ray Burst (GRB) data is used to constraint the NC parameters. In Chapter 4, the collapse of thin null shells are considered, as the extension of two Schwarzschild solutions with different masses. The quantization of this setup is studied and the differences between the commutative algebra and the deformed one are analysed. In Chapter 5, the quantization of FRW-type metric is performed, using WdW equation in the context of Hořava-Lifshitz gravity. A synthesis of our conclusions is presented in Chapter 6.

1.1 Quantum Mechanics in Phase-space

1.1.1 Phase-space formulation

Quantum mechanics (QM) is most often formulated using the Hilbert space approach, with operators acting on wave function to compute physically relevant quantities. However, other formulations, which are equivalent to this one, exist, such as path integral formulation or phase-space (PS) formulation. While the former is mainly used in Quantum Field Theory, the latter is useful in the description of transportation processes in phase-space, and also in quantum optics, nuclear physics, in the study of the classical limit of mesoscopic systems and the transition from quantum mechanics to classical statistical physics. The PS formulation is also used in information theory as a main tool to understand information transmission phenomena.

This formulation of QM has its origin with the work of Wigner and Weyl [28, 29]. The former defined the Wigner quasi-probability distribution and the latter proved the equivalence between operators in Hilbert space and functions in phase-space. These concepts were later put together by Groenwold and Moyal [30, 31], who also defined the Moyal star-product, which is central for the theory since it defines the equivalent of operator product in Hilbert space.

One of the prime features of this quantization approach is the Wigner quasi-probability distribution, which, for a 2n-dimensional phase-space, is defined as:

$$f(x,p) = \frac{1}{(2\pi\hbar)^n} \int \left\langle x - \frac{y}{2} \middle| \rho \middle| x + \frac{y}{2} \right\rangle e^{ip\cdot y/\hbar} d^n y, \tag{1.1}$$

where ρ is the density matrix for the system. For a pure state in a 2-dimensional phase-space this reduces to:

$$f(x,p) = \frac{1}{2\pi\hbar} \int \psi^* \left(x - \frac{y}{2}\right) \psi \left(x + \frac{y}{2}\right) e^{ipy/\hbar} dy. \tag{1.2}$$

This definition of the Wigner function (WF) relies on the wave function of the system at hand. However, this only serves the purpose of connection with the Hilbert space approach to quantization. As will be seen, there is no need for the input of wave functions to compute the Wigner function; in fact, just as the wave functions are determined by a differential equation, so will the Wigner function. The main goal of the PS formulation is to compute the WF for that system. It then allows for the computation of all relevant physical quantities. This function has a number of useful interesting properties, that hint at its quantum nature as a generalization of some classical features. First, one can see that taking the limit $\hbar \to 0$ the WF reduces to the probability density in the coordinate space x and a Dirac delta in p-space, meaning it would describe a classical state with a very well defined momenta. In reverse, one can also define the WF by:

$$f(x,p) = \frac{1}{2\pi\hbar} \int \varphi^* \left(p - \frac{q}{2}\right) \varphi \left(p + \frac{q}{2}\right) e^{iqx/\hbar} dq, \tag{1.3}$$

where φ represents the Fourier transform of ψ. Taking the same limit in Eq. (1.3) leads to a probability density in momentum space with a well defined position. Additionally, this function is bounded from both above and below, as given by:

$$-\frac{2}{\hbar} \le f(x,p) \le \frac{2}{\hbar}, \tag{1.4}$$

a bound that disappears in the classical limit. This also reveals that, in most scenarios, the WF is negative in some regions of PS, thus a direct interpretation as a probability distribution for states in this space is not suitable. In fact, the only pure state WF that is non-negative everywhere on the phase-space is the Gaussian WF (insert citation). Nonetheless, all physical quantities computed from it are proper probability distributions. Additionally, the Wigner function is also real everywhere and $\int f(x,p)dx = \sigma(p) = |\varphi(p)|^2$ and $\int f(x,p)dp = \rho(x) = |\psi(x)|^2$ have the properties of marginal probability distributions for p and x, respectively (the second equality is valid for pure states only).

Now, instead of pursuing with the connection between the two formulations, let us first define completely the PS one. As previously stated, the description of quantum phenomena using this approach relies solely on complex valued function, i.e. c-functions, and in a noncommutative, associative, pseudo-differential product of functions called Moyal star-product (or $\star$-product). Given two functions, $a(z), b(z) \in \mathbb{C}^n$ ($z = (x,p)$) (for more notation see Notation and Conventions), this product can be defined as:

$$a(z) \star b(z) := a(z) e^{\frac{i\hbar}{2} \overleftarrow{\partial}_{z_\alpha} J_{\alpha\beta} \overrightarrow{\partial}_{z_\beta}} b(z), \tag{1.5}$$

where $J_{\alpha\beta}$ are the components of the standard symplectic matrix J:

$$J = \begin{pmatrix} 0 & 1_n \\ -1_n & 0 \end{pmatrix}, \tag{1.6}$$

where 1_n is the n-dimensional identity matrix. Alternatively, in integral form, this product is given by:

$$a(z) \star b(z) = \frac{1}{(\pi\hbar)^{2n}} \int a(z')b(z'')e^{(-2i/\hbar)[(z-z')J(z''-z)^\dagger]}\, dz'dz''. \tag{1.7}$$

Given this product, the dynamical evolution of the WF is given by the Moyal equation, which reads:

$$\frac{\partial f}{\partial t} = \{\{H(z), f(z)\}\}, \tag{1.8}$$

where the double curly brackets represent the so called Moyal bracket, which is defined as:

$$\{\{a, b\}\} := \frac{1}{i\hbar}\left(a \star b - b \star a\right). \tag{1.9}$$

This equation resembles Liouville's theorem from Classical Mechanics and in fact is an extension of it in the sense that, when taking $\hbar \to 0$ one recovers that equation. For static problems, this equation reduces to a simpler one, much in the same way the time dependent Schrödinger equation reduces to the time independent one. Thus it is written as:

$$H(z) \star f(z) = f(z) \star H(z) = Ef(z), \tag{1.10}$$

a $\star$-genvalue equation that determines the spectrum of the system, where E is the energy of a given state. Solving these equations amounts for the complete characterization of the Wigner functions for a system. Having the WF for some state is only part of the discussion, since one must also be able to compute expectation values for observables. Given an observable $g = g(z)$, its expectation value is computed through:

$$\langle g \rangle = \int f(z)g(z)\mathrm{d}^n z, \tag{1.11}$$

which can be interpreted as a phase-space average, weighted by the Wigner function. This result also allows for the interpretation of the WF as the distribution that defines the integration measure on phase-space. Given that these equations allow us for completely solving the system, when provided its Hamiltonian, one must check if this approach is indeed equivalent to the mainstream formulations. For this, one must define the so called Weyl transform.

This is a one-to-one map from operators on Hilbert space, $L^2(\mathbb{R}^n)$, and c-functions on phase-space. It is defined as:

$$W[\hat{G}(\hat{z})] := g(z) = \frac{1}{(2\pi\hbar)^n} \int \left\langle x - \frac{y}{2} \middle| \hat{G} \middle| x + \frac{y}{2} \right\rangle e^{ip\cdot y/\hbar}\, d^n y. \tag{1.12}$$

The phase-space function $g(z)$ is real iff $\hat{G}$ is self-adjoint. For polynomial operators, this function is usually obtained by making $\hat{p} \to p$ and $\hat{x} \to x$. Interestingly, from the above formula, and considering that $W[\hat{I}] = 1$, one obtains that, for normalized states:

$$1 = \int f(x,p)d^n z \tag{1.13}$$

Being one-to-one, this admits an inverse, mapping each c-function to some Hilbert space operator as:

$$\hat{G}(\hat{z}) = \int g(x,p)e^{ia(\hat{P}-p)+ib(\hat{X}-x)}\, da\, db\, dx\, dp. \tag{1.14}$$

Besides connecting both formulations of QM, this particular transform has the compatible with the $\star$-product, in the sense that,

$$W[\hat{G}\hat{I}] = W[\hat{G}] \star W[\hat{I}] = g(z) \star i(z). \tag{1.15}$$

Thus, the Moyal $\star$-product is in isomorphism with operator product in the Hilbert space. Therefore, by computing some Weyl transforms of simple operators, one is able to get the WF of more complex operators by breaking them into simpler parts, e.g. using a Taylor expansion. With the above results the trace equation can be written as:

$$tr\left(\hat{G}\right) = \int dz g(z), \tag{1.16}$$

which, in turn, allows for the expectation value equation, Eq. (1.11), as $tr\left(\hat{\rho}\hat{G}\right)$.

With these tools, we are able to see that the energy in the time-independent Moyal equation matches the energy definition of the time independent Schrödinger equation. This can be seen in the following:

$$
\begin{aligned}
H(z) \star f(z) &= \frac{1}{(\pi\hbar)^{2n}} \int H(z')f(z'')e^{(-2i/\hbar)[(z-z')J(z''-z)^\dagger]}\, dz'dz'' \\
&= \frac{1}{(\pi\hbar)^{2n}} \int H(z')f(z'')e^{(-2i/\hbar)[p\cdot(x'-x'')+p'\cdot(x''-x)+p''\cdot(x-x')]}\, dx'dp'dx''dp'' \\
&= \frac{1}{(\pi\hbar)^{2n}}\frac{1}{(2\pi\hbar)^{2n}} \int e^{ip'\cdot y/\hbar}\left\langle x' - \frac{y}{2}\middle|\hat{H}\middle|x' + \frac{y}{2}\right\rangle \psi_n^*\left(x'' + \frac{y'}{2}\right)\psi\left(x'' - \frac{y'}{2}\right)e^{ip''\cdot y'/\hbar} \times \\
&\quad \times e^{(-2i/\hbar)[p\cdot(x'-x'')+p'\cdot(x''-x)+p''\cdot(x-x')]}\, dx'dp'dx''dp''dy dy' = \\
&= \frac{1}{(\pi\hbar)^{2n}}\frac{1}{(2\pi\hbar)^{2n}} \int \left\langle x' - \frac{y}{2}\middle|\hat{H}\middle|x' + \frac{y}{2}\right\rangle \psi_n^*\left(x'' + \frac{y'}{2}\right)\psi\left(x'' - \frac{y'}{2}\right) \times
\end{aligned}
$$

$$\times \, e^{ip'\cdot(y-2(x''-x))/\hbar}\,e^{ip\cdot(y'-2(x-x'))/\hbar}\,e^{-2ip\cdot(x'-x'')/\hbar}\,dx'dp'dx''dp''dydy'$$

$$= \frac{1}{(\pi\hbar)^{2n}} \int \left\langle x' - \frac{y}{2} \left| \hat{H} \right| x' + \frac{y}{2} \right\rangle \psi_n^* \left(x'' + \frac{y'}{2} \right) \psi \left(x'' - \frac{y'}{2} \right) \times$$

$$\times \, \delta^n(y - 2(x'' - x))\delta^n(y' - 2(x - x'))e^{-2ip\cdot(x'-x'')/\hbar}\,dx'dx''dydy'$$

$$= \frac{1}{(\pi\hbar)^{2n}} \int \left\langle x' - x'' + x \left| \hat{H} \right| x' + x'' - x \right\rangle \psi^* \left(x'' + x - x' \right) \psi \left(x'' - x + x' \right) \times$$

$$\times \, e^{-2ip\cdot(x'-x'')/\hbar}\,dx'dx''$$

$$(1.17)$$

Here we used the definitions of the Weyl transform of the Hamiltonian operator, $\hat{H}$, as well as the definition of the Wigner function for a given state ψ_n. The integral representation of the Dirac-delta distribution was used when integrating in the momenta p' and p''. From here, it will be useful to make use of the closure relation of a complete set of states for the system. The states chosen will be proper states of the Hamiltonian operator, for convenience. Additionally, one will assume these states to be discrete and nondegenerate for simplicity. If this is not the case, the generalization of the remainder of this calculation is simple and the result is unchanged. Then:

$$H(z) \star f(z) = \frac{1}{(\pi\hbar)^{2n}} \sum_{k,l} \int \left\langle x' - x'' + x \,\middle|\, \psi_k \right\rangle \left\langle \psi_k \middle| \hat{H} \middle| \psi_l \right\rangle \left\langle \psi_l \,\middle|\, x' + x'' - x \right\rangle \times$$

$$\times \, \psi_n^* \left(x'' + x - x' \right) \psi \left(x'' - x + x' \right) e^{-2ip\cdot(x'-x'')/\hbar}\,dx'dx''$$

$$= \frac{1}{(\pi\hbar)^{2n}} \sum_l E_l \int \psi_l(x' - x'' + x)\psi_l^*(x' + x'' - x)\psi_n^* \left(x'' + x - x' \right) \psi \left(x'' - x + x' \right) \times$$

$$\times \, e^{-2ip\cdot(x'-x'')/\hbar}\,dx'dx''$$

$$= \frac{1}{2^n(\pi\hbar)^{2n}} \sum_l E_l \int \psi_l(x_- + x)\psi_l^*(x_+ - x)\psi_n^* \left(x - x_- \right) \psi_n \left(x_+ - x \right) e^{-2ip\cdot x_-/\hbar}\,dx_-dx_+$$

$$= \frac{1}{(2\pi\hbar)^{2n}} \sum_l E_l \int \delta_{l,n}\psi_l(x_- + x)\psi_n^* \left(x - x_- \right) e^{-2ip\cdot x_-/\hbar}\,dx_-$$

$$= E_n \frac{1}{(2\pi\hbar)^{2n}} \int \psi_n \left(x + \frac{\bar{x}}{2} \right) \psi_n^* \left(x - \frac{\bar{x}}{2} \right) e^{-ip\cdot\bar{x}/\hbar}\,d\bar{x}$$

$$= E_n f(z).$$

This ensures that the values of E_n in the Moyal equation have the same interpretation as in the Schrödinger equation: the energy of the state represented by f_n and ψ_n, respectively. Another crucial issue of the phase-space formulation is the computation of expected values for observables, Eq. (1.11). This relationship can now also be proven to be equivalent to the same quantity in the usual approach to QM following a similar procedure. Given a state with WF $f_n(z)$ and an observable $g(z)$, by Eq. (1.11) we have:

$$
\begin{aligned}
\langle g \rangle_n &= \int g(z) f(z)\, \mathrm{d}^n z \\
&= \frac{1}{(2\pi\hbar)^{2n}} \int \left\langle x - \tfrac{y}{2} \middle| \hat{G} \middle| x + \tfrac{y}{2} \right\rangle e^{\mathrm{i} p \cdot y} \psi_n^* \left(x - \tfrac{\overline{y}}{2} \right) \psi_n \left(x + \tfrac{\overline{y}}{2} \right) e^{-\mathrm{i} p \cdot \overline{y}}\, \mathrm{d}^n x\, \mathrm{d}^n p\, \mathrm{d}^n y\, \mathrm{d}^n \overline{y} \\
&= \frac{1}{(2\pi\hbar)^{n}} \int \left\langle x - \tfrac{y}{2} \middle| \hat{G} \middle| x + \tfrac{y}{2} \right\rangle \psi_n^* \left(x - \tfrac{y}{2} \right) \psi_n \left(x + \tfrac{y}{2} \right)\, \mathrm{d}^n x\, \mathrm{d}^n y \\
&= \frac{1}{(\pi\hbar)^{n}} \sum_{k,l} \int \langle \psi_k | \hat{G} | \psi_l \rangle\, \psi_k (x - y)\, \psi_l^* (x + y)\, \psi_n^* (x - y)\, \psi_n (x + y)\, \mathrm{d}^n x\, \mathrm{d}^n y \\
&= \frac{1}{(2\pi\hbar)^{n}} \sum_{k,l} \int \langle \psi_k | \hat{G} | \psi_l \rangle\, \psi_k (x_-)\, \psi_l^* (x_+)\, \psi_n^* (x_-)\, \psi_n (x_+)\, \mathrm{d}^n x_-\, \mathrm{d}^n x_+ \\
&= \frac{1}{(2\pi\hbar)^{n}} \sum_{l} \int \langle \psi_n | \hat{G} | \psi_l \rangle\, \psi_l^* (x_+)\, \psi_n (x_+)\, \mathrm{d}^n x_+ \\
&= \frac{1}{(2\pi\hbar)^{n}} \langle \psi_n | \hat{G} | \psi_n \rangle = \langle \hat{G} \rangle_n
\end{aligned}
\tag{1.18}
$$

This verification completes the correspondence with the standard formulation of QM. The PS formulation is independent from the Schrödinger formulation and it has the same interpretation for the expected values of observables. Therefore, one should look at phase-space formulation on its own, independent and equivalent to other formulations of quantum mechanics.

After this brief introduction to the phase-space formalism of QM, one should analyse its features, in order to acquire a deeper understanding. In particular, it is noticeable that there are no operators involved in this formalism. In fact, the only feature that differs from a classical treatment is the existence of the Moyal product. This is a key observation, as this $\star$-product is responsible for encoding all quantum features. This product can be regarded as a quantum correction to classical physics, in the sense that the series expansion that defines it, i.e. Eq. (1.5), gets smaller contributions from higher order terms and when taking $\hbar \to 0$ one is left with the classical setting.

1.1.2 Noncommutative Quantum Mechanics

Noncommutative Quantum Mechanics (NCQM) is an extension of Quantum Mechanics that admits extra commutation relations that would otherwise vanish. In particular, it consists of extending the position and momentum operators no to commute with themselves (for different spatial directions). Mathematically this is translated to a deformation of the standard

Heisenberg-Weyl (HW) algebra [15]:

$$
\begin{aligned}
\left[\hat{x}_i, \hat{p}_j\right] &= i\hbar\delta_{ij} & \longrightarrow && \left[\hat{q}_i, \hat{k}_j\right] &= i\hbar\delta_{ij} \\
\left[\hat{x}_i, \hat{x}_j\right] &= 0 & \longrightarrow && \left[\hat{q}_i, \hat{q}_j\right] &= i\theta_{ij} \\
\left[\hat{p}_i, \hat{p}_j\right] &= 0 & \longrightarrow && \left[\hat{k}_i, \hat{k}_j\right] &= i\eta_{ij},
\end{aligned}
\tag{1.19}
$$

where θ_{ij} and η_{ij} are rank $\binom{0}{2}$ antisymetric tensor. The components of these tensors are taken to be constant, and one may choose to write $\theta_{ij} = \theta\epsilon_{ij}$ and $\eta_{ij} = \eta\epsilon_{ij}$. Here ϵ_{ij} is used according to the conventions set in the Notation and Conventions section.

When working within this framework, two possibilities arise: either one tries to correct the usual Hamiltonians, by adding counterterms that cancel the NC effects, or one assumes that the Hamiltonian is correct and tries to infer the consequences of the extended commutation relations. This second approach will be taken henceforth. We assume that the correct Hamiltonians have been used, but the NC effects are too small to be noticed. Thus, in order to study some system in this setup one takes $\hat{H}(\hat{x}, \hat{p}) \to \hat{H}\left(\hat{q}, \hat{k}\right)$, i.e., the form of $\hat{H}$ remains the same, only the position and momenta are now assumed to obey the modified HW algebra. As an example, the NC harmonic oscillator Hamiltonian is then written as:

$$
\hat{H}_{HO} = \frac{\hat{k}^2}{2m} + \omega^2 \hat{q}^2.
\tag{1.20}
$$

Usually, one can work out the spectrum of a given system in several different ways. One of them is using the position (or momentum) representation of $\hat{x}$ and $\hat{p}$ and the problem reduces to a partial differential equation problem. Another approach is to define ladder operators to construct the spectrum by acting on the vacuum state. In whatever way, these techniques are dependent on the HW algebra: in the first, the differential operators directly obey this algebra and in the second the algebra of the ladder operators and the HW algebra define the form of these operators. However, when one deforms the HW algebra, this approaches must also change. For one, the position and momentum representations no longer work in the simple way one is used to; and the relation between position and momentum and the ladder operators must also change. One way to solve this is by using the so called Seiberg-Witten (SW) map [13] – or Darboux map, as its called in symplectic geometry; this will be covered in the next section. This map defines a noncanonical transformation in the phase-space that allows us to relate (x, p) to (q, k) in a way that preserves the commutation relations of each variable set. It may be defined as:

$$
\begin{aligned}
D: \quad & \mathbb{R}^n \longrightarrow \mathbb{R}^n \\
& (q, k) \longrightarrow (x, p) = D(q, k)
\end{aligned}
\tag{1.21}
$$

In this work, we use a noncanonical linear map, given by [13, 15]

$$\hat{q}^i = \lambda \hat{x}^i - \frac{\theta}{2\lambda\hbar} \epsilon^{ij} \hat{p}_j, \quad \hat{k}_i = \mu \hat{p}_i + \frac{\eta}{2\mu\hbar} \epsilon_{ij} \hat{x}^j, \tag{1.22}$$

which is invertible and has unit Jacobian determinant provided the parameters λ and μ are constrained by the relationship

$$\frac{\theta\eta}{4\hbar^2} = \lambda\mu(\lambda\mu - 1), \tag{1.23}$$

so to give

$$\begin{aligned}
\hat{x}^i &= \mu \left(1 - \frac{\theta\eta}{\hbar^2}\right)^{-1/2} \left(\hat{q}^i + \frac{\theta}{2\lambda\mu\hbar} \epsilon^{ij} \hat{k}_j\right), \\
\hat{p}_i &= \lambda \left(1 - \frac{\theta\eta}{\hbar^2}\right)^{-1/2} \left(\hat{k}_i - \frac{\eta}{2\lambda\mu\hbar} \epsilon_{ij} \hat{q}_j\right),
\end{aligned} \tag{1.24}$$

with $\epsilon_{ij} = -\epsilon_{ji}$, $\epsilon_{ij} = \pm 1$, $\theta\eta \lesssim \hbar^2$, and the corresponding Jacobian reading

$$\frac{\partial(q,k)}{\partial(x,p)} = \sqrt{\det(\Omega)} = 1 - \frac{\theta\eta}{\hbar^2}. \tag{1.25}$$

By completeness, in its most general form, the SW map can be expressed as [15]:

$$\hat{x}_i = A_{ij}\hat{q}_j + B_{ij}\hat{k}_j, \qquad \hat{p}_i = C_{ij}\hat{q}_j + D_{ij}\hat{k}_j, \tag{1.26}$$

where $\mathbf{A}, \mathbf{B}, \mathbf{C}, \mathbf{D}$ are real constant matrix solutions of the equations

$$\mathbf{AD}^T - \mathbf{BC}^T = \mathbf{I}_{n\times n} \qquad \mathbf{AB}^T - \mathbf{BA}^T = \frac{1}{\hbar}\mathbf{\Theta} \qquad \mathbf{CD}^T - \mathbf{DC}^T = \frac{1}{\hbar}\mathbf{N}, \tag{1.27}$$

where the superscript "T" stands for matrix transposition and A_{ij}, B_{ij}, C_{ij}, D_{ij}, θ_{ij}, η_{ij} are the entries of the matrices $\mathbf{A}$, $\mathbf{B}$, $\mathbf{C}$, $\mathbf{D}$, $\mathbf{\Theta}$, $\mathbf{N}$, respectively.

The above linear transformations imply that the NC algebra expressed by the relations from Eq. (1.22) admits a representation in terms of the Hilbert space of the ordinary QM which provides a self-contained phase-space formulation of the NC QM. This particular issue will be dealt with in Subsection 1.1.4.

This map allows for the mapping of one set of operators into the other, keeping the algebra of operators consistent. This can now be used in the Hamiltonian, in order to write a NC system in terms of commutative operators, thus being able to use the standard position (or momentum) representation of these operators. As an example, setting $\lambda, \mu = 1$, using the

HO above, one gets:

$$\hat{H}_{HO} = \frac{1}{2m}\left(\hat{p}_i + \frac{\eta}{2\hbar}\epsilon^{ij}\hat{x}_j\right)^2 + \omega^2\left(\hat{x}_i - \frac{\theta}{2\hbar}\epsilon^{ij}\hat{p}_j\right)^2 \tag{1.28}$$

which, for 2 dimensions, becomes:

$$\begin{aligned}
\hat{H}_{HO} &= \frac{\hat{p}^2}{2m} + \frac{\eta^2}{8m\hbar^2}\hat{x}^2 + \frac{\eta}{2m\hbar}\left(\hat{p}_1\hat{x}_2 - \hat{p}_2\hat{x}_1\right) + \omega^2\hat{x}^2 + \frac{\omega^2\theta^2}{4\hbar^2}\hat{p}^2 - \frac{\omega^2\theta}{\hbar}\left(\hat{p}_1\hat{x}_2 - \hat{p}_2\hat{x}_1\right) \\
&= \left(\frac{1}{2m} + \frac{\omega^2\theta^2}{4\hbar^2}\right)\hat{p}^2 + \left(\omega^2 + \frac{\eta^2}{8m\hbar^2}\right)\hat{x}^2 + \left(\frac{\eta}{2m\hbar} - \frac{\omega^2\theta}{\hbar}\right)\left(\hat{p}_1\hat{x}_2 - \hat{p}_2\hat{x}_1\right).
\end{aligned} \tag{1.29}$$

Thus, the harmonic oscillator gains an extra term, akin to an applied magnetic field in the direction perpendicular to the $x - y$ plane. One would now be able to apply standard techniques to this NC Hamiltonian written with commutative operators. Consequently, with this map there is an isomorphism between Hamiltonians with NC operators and the ones written in terms of commutative ones. This is the fact we use to solve these problems. In spite of this, for certain applications, such as quantum cloning and teleportation, one might be interested in formulating NCQM in phase-space. This shall be done in the next sections, as we dive into symplectic geometry to gain some insight in the geometrical properties of phase-space.

1.1.3 Symplectic Geometry

Symplectic geometry is the study of differential manifolds equipped with a closed, nondegenerate 2-form. It is, therefore, a branch of differential geometry, in the same way that Riemannian geometry is the branch of this subject that studies differential manifolds equipped with a nondegenerate, symmetric 2-tensor. Since both are branches of the same general subject, the language of symplectic and Riemannian geometry is the same. However, due to the nature of their intrinsic structures, their properties are quite different.

Consider the symplectic structure in more detail. Given a vector field V and $v, u, z \in V$, one can define it as a differential 2-form, ω, with the properties:

$$\begin{aligned}
&\omega\left(v + z, u\right) = \omega\left(v, u\right) + \omega\left(z, u\right), \\
&\omega\left(v, u + z\right) = \omega\left(v, u\right) + \omega\left(v, z\right), \\
&\omega\left(v, u\right) = -\omega\left(u, v\right), \\
&\text{If } \omega(v, u) = 0 \text{ for all } u \in V, \text{ then } v = 0.
\end{aligned} \tag{1.30}$$

A vector field equipped with such a form is called a symplectic vector field and it is denoted as the pair (V, ω). One can then regard ω as a map:

$$\omega : V \times V \rightarrow F, \tag{1.31}$$

where the field F may be $\mathbb{R}$ or $\mathbb{C}$. Alternatively, the same form may be regarded as a map to the dual space of V, V^*, the space of linear functionals of V, i.e.,

$$\begin{aligned}
\omega : V &\to V^* \\
v &\to \omega(v, \cdot) := v^*,
\end{aligned}$$
(1.32)

since $v^*(u) \in \mathbb{R}$, for $u \in V$. Consequently, ω^{-1} is easily seen as a map from V^* to V. This bypasses the ambiguity of defining the inverse from $\mathbb{R}$ to $V \times V$, which would be ill-defined since ω is not injective when regarded in the sense of Eq. (1.31).

Given these properties, it is possible to construct several different symplectic forms. There is, however, a standard symplectic form, $\boldsymbol{J}$, defined for $\mathbb{R}^{2n}$, similar to the Euclidean metric in Riemannian geometry, but with a much more important role here, given by:

$$\boldsymbol{J} = \begin{pmatrix} 0 & \mathbf{1}_n \\ -\mathbf{1}_n & 0 \end{pmatrix},$$
(1.33)

where $\mathbf{1}_n$ is the n-dimensional identity matrix. The importance of this object will become clear shortly. One of the differentiating properties of symplectic geometry is that symplectic vector fields must be of even dimension. This can be seen in the following. Consider a n-dimensional vector field and the matrix representation of the bilinear form ω, Ω. On account of ω being antisymetric, $\Omega = -\Omega^t$, hence $\det(\Omega) = (-1)^n \det(\Omega^t) = (-1)^n \det(\Omega)$. As ω is non degenerate, $\det(\Omega) \neq 0$, thus $1 = (-1)^n$. When working over $\mathbb{R}$ or $\mathbb{C}$[1], this implies that $n = 2m$, and consequently V has even dimension. Then, we can always choose coordinates for symplectic vector fields as (x^i, y_i) with $i = 1, \cdots, n$ and write the symplectic form in Eq. (1.33) in differential form language:

$$J = \mathrm{d}x^i \wedge \mathrm{d}y_i.$$
(1.34)

One of the most remarkable results in symplectic geometry, Darboux's Theorem, states that, it is always possible to find a set of coordinates $(\tilde{x}_i, \tilde{y}_i)$, called canonical coordinates, such that the symplectic form becomes as in Eq. (1.33) in an entire neighborhood of any point P. This result may seem familiar from Riemannian geometry, where one is always able to choose a coordinate system to make the metric tensor take the standard form at any point. The important difference is that Darboux's Theorem states that the symplectic form takes the canonical form in an entire neighborhood of that point. This implies that, while in Riemannian geometry the derivatives of geometrical quantities are still non-vanishing (at least

---

[1] For fields with characteristic 2 this does not hold, since $(-1)^n = 1$ for all values of n. Since the characteristic of either $\mathbb{R}$ or $\mathbb{C}$ is 0 this doesn't pose a problem. A proof for $\mathrm{char}(F) = 2$ is out of the scope of this work.

second derivatives), in symplectic vector spaces all quantities computed from ω are indistinguishable from the same quantities computed from J. This, in turn, implies that there are no geometrical invariants in symplectic geometry. This means that given two symplectic vector spaces, no local quantity can be computed that distinguishes those entities. This in marked contrast with Riemannian geometry where quantities such as the Riemann curvature tensor can be computed. Consequently, only global invariants can be computed in symplectic vector spaces. The most useful example with respect to physics is the quantity:

$$A = \int_S \omega, \tag{1.35}$$

i.e., the area of the surface S. This can be used in Hamiltonian mechanics, as in dynamical systems this quantity is invariant under temporal evolution. These concepts introduced in vector spaces generalized in a simple way to manifolds as a consequence of the tangent space on a point P, $T_P M$, of the manifold being a vector space. Then, one can equip this space with a symplectic form and define it as a map from the tangent space to the cotangent space, $T_P^* M$, i.e.,

$$\begin{aligned}
\omega_P &: T_P M \to T_P^* M \\
v &\to \omega_P(v, \cdot) := v^*,
\end{aligned} \tag{1.36}$$

for $v \in T_P M$. Alternatively, $\omega_P : T_P M \times T_P M \to \mathbb{R}$ so $\omega_P \in \Lambda^2(T_P M)$. Consider now a differential manifold M and a differential 2-form on M, $\omega \in \Omega^2(M)$. Then, M is called symplectic if $\forall P \in M$, $\omega_P := \omega|_P$ is a symplectic form on $T_P M$ and ω is closed, i.e., $d\omega = 0$. A direct consequence of this definition is that the dimension of the tangent space must be even and since $\dim(T_P M) = \dim(M)$, the manifold's dimension must also be even.

The main link of symplectic geometry and physics lies in the fact that the phase-space can be regarded as a symplectic manifold. Thus, classical Hamiltonian mechanics may be viewed from this perspective and the formal description of physical system may be cast in the language of differential geometry and symplectic forms. But the question remains, why symplectic forms and not metrics tensors (which would lead to Riemannian geometry)? The main factor is the fact that the Poisson brackets are related to the symplectic form in the as:

$$\{f, g\} = \omega(X_f, X_g), \tag{1.37}$$

where $X_{f/g}$ are the vector fields associated with the corresponding function. These are defined through the relation:

$$\iota_{X_f} \omega = -df, \tag{1.38}$$

and ι_X is the interior product of a vector field and a differential form, $\iota_X : \Omega^p(M) \to$

$\Omega^{p-1}(M)$, defined by $(\iota_X\omega)(X_1,\ldots,X_n) = \omega(X,X_1,\ldots,X_n)$. If one is interested in the matrix representation of the symplectic the components of this form are given by the poisson brackets of the coordinates. Let us consider coordinates $z^i = (x^i, p_i)$, as introduced in the Notation and Conventions. Then, Eq. (1.38) in component form becomes:

$$\omega_{ij}X^i_f = -\partial_j f \Leftrightarrow X^i_f = -\omega^{ij}\partial_j, \tag{1.39}$$

where $\omega^{ij} = (\omega)^{-1}_{ij}$ represents the components of the inverse of the symplectic form (see Eq (1.32) and ensuing discussion). Then, it is possible to rewrite Eq. (1.37) as:

$$\{f,g\} = \omega_{ij}X^i_f X^j_g = \omega_{ij}\omega^{ik}\omega^{jl}\partial_k f \partial_l g = \omega^{kl}\partial_k f \partial_l g. \tag{1.40}$$

In particular, if one considers $f(z) = z^i$ and $g(z) = z^j$, the previous equation reduces to:

$$\{z^i, z^j\} = \omega^{ij}. \tag{1.41}$$

The statement is then that the Poisson brackets of the coordinates are directly tied with the symplectic form, consequently making the framework of symplectic geometry fitting for describing Hamiltonian dynamics. In addition, seeing that, in the standard quantization procedure, the algebra of position and momentum operators stems from the promotion of Poisson brackets to commutators, one can also consider the relation:

$$[f,g] = i\hbar\omega(X_f, X_g), \tag{1.42}$$

which in turn implies that in QM,

$$[\hat{z}^i, \hat{z}^j] = i\hbar\omega^{ij}. \tag{1.43}$$

Accordingly, the algebra of operators will determine the symplectic structure of the phase-space. It should now be clear the reason why the symplectic form is at the heart of the definition of the Moyal $\star$-product, Eq. (1.5), being that the commutators determine this product of functions and they can, in turn, be determined by the symplectic form.

The connection between ω and the commutators of $\hat{x}$ and $\hat{p}$, together with Darboux's theorem also implies that, for any algebra with a finite number of elements (i.e., forming a finite dimensional vector space), it is always possible to find a coordinate transformation such that the new commutation relations are those given by the HW algebra.

1.1.4 Phase-space formalism for NCQM

Having introduced the symplectic geometry and its connection of QM and phase-space formalism, we can now proceed to apply this formalism to NCQM. Upon the introduction of this formalism for QM one is able to notice that all quantum features arise as the result of the Moyal $\star$-product, as a $\hbar$-deformation of the usual product of functions, which led to a deformed Poisson bracket, namely the Moyal bracket. This was the basis of the formalism, and the kinematic and dynamic equations of QM were written with this product. Therefore, taking into account that the $\hbar$-deformation of the product is a consequence of the algebra of those operators in the Hilbert space formulation, one can expect that a PS version of NCQM will have some deformation of the product based on the additional commutation relations introduced in Eq. (1.19). Moreover, one could anticipate that this change is to be related to the symplectic structure of phase-space, because the symplectic form J is present in the definition of the Moyal product, Eq. (1.5). In fact, the Wigner-Weyl or phase-space formulation of the NC QM is a particular case of the deformation quantization method, which is a procedure alternative to the canonical operator quantization or path integral formalism. This formalism can be applied to classical systems and is consistent for non-flat or sympletic manifolds, as is the case of phase-space. This leads to a similar structure to statistical mechanics, where observables are represented by phase-space functions and the Wigner function acts as a quasi-probability distribution on this space. The corrections to classical behaviour are then encoded in the associative, noncommutative $\star$-product defined on the manifold. Different deformations of the Poisson brackets then lead to different products, that convey the information of each deformation. When the deformation parameters, $\hbar$ on standard QM and θ and η in NC QM, are taken to zero, the ordinary function product on the phase-space is recovered, and classical results are recovered.

However, it must guaranteed that it is possible to define a product with the same properties of the Moyal $\star$-product – associativity, noncommutativity and that it obeys property Eq. (1.15) for a given NC WW transform.

Following the above reasoning, one can start by defining the deformed $\star$-product by its action on phase-space functions as [15]:

$$a(\tilde{z}) \star_{NC} b(\tilde{z}) := a(\tilde{z}) e^{\frac{i\hbar}{2} \overleftarrow{\partial}_{\tilde{z}_\alpha} \Omega_{\alpha\beta} \overrightarrow{\partial}_{\tilde{z}_\beta}} b(\tilde{z}), \tag{1.44}$$

where now the Ω matrix is given by:

$$\Omega = \begin{pmatrix} \frac{1}{\hbar}\Theta & 1_n \\ -1_n & \frac{1}{\hbar}N \end{pmatrix}, \tag{1.45}$$

with Θ and N being the matrices with elements θ_{ij} and η_{ij}. It is easy to see that:

$$[\tilde{z}_i, \tilde{z}_j] = i\hbar\Omega_{ij}. \tag{1.46}$$

Thus, considering Eq. (1.43), $\Omega_{ij} = \omega^{ij} = (\omega^{-1})_{ij}$ and so Ω is a matrix representation of the inverse of the symplectic form for the NC phase-space.

In fact, this $\star$-product is a suitable deformation of the one defined for the phase-space QM formulation as Eq. (1.5) is recovered in the limit where the NC parameters, θ and η, vanish. Furthermore, since the Ω matrix, Eq. (1.45), can be decomposed,

$$\Omega = J + \begin{pmatrix} \frac{i}{\hbar}\Theta & 0 \\ 0 & 0 \end{pmatrix} + \begin{pmatrix} 0 & 0 \\ 0 & \frac{1}{\hbar}N \end{pmatrix}, \tag{1.47}$$

the Moyal $\star$-product can itself be written as to cast Eq. (1.44)

$$a(\tilde{z}) \star_{NC} b(\tilde{z}) = a(\tilde{z}) \star_{\hbar} \star_{\theta} \star_{\eta} b(\tilde{z}). \tag{1.48}$$

Here, the three products are defined by the three matrices that form the decomposition of Ω in Eq. (1.45), and reduce to:

$$a(\tilde{z}) \star_{\hbar} b(\tilde{z}) := a(\tilde{z}) e^{\frac{i\hbar}{2} \overleftarrow{\partial}_{z_\alpha} J_{\alpha\beta} \overrightarrow{\partial}_{z_\beta}} b(\tilde{z}), \tag{1.49a}$$

$$a(\tilde{z}) \star_{\theta} b(\tilde{z}) := a(\tilde{z}) e^{\frac{i}{2} \overleftarrow{\partial}_{z_\alpha} \theta_{\alpha\beta} \overrightarrow{\partial}_{z_\beta}} b(\tilde{z}), \tag{1.49b}$$

$$a(\tilde{z}) \star_{\eta} b(\tilde{z}) := a(\tilde{z}) e^{\frac{i}{2} \overleftarrow{\partial}_{z_\alpha} \eta_{\alpha\beta} \overrightarrow{\partial}_{z_\beta}} b(\tilde{z}), \tag{1.49c}$$

However, in order to make sure that the PS formulation of NCQM matches the standard formulation, one must define a generalized Wigner-Weyl transform, $W^{NC}[\hat{A}]$, for some operator $\hat{A}$ in the theory with the deformed HW algebra, such that there is a one-to-one map from operators to phase-space functions. Moreover, there also needs to be an equation of motion that matches the Schrödinger equation for the new algebra. For this purpose, one can take advantage of the unique characteristic of symplectic geometry: Darboux's theorem. By making use of this theorem, one is always able to find coordinates such that the algebra becomes the standard HW one and then one can use the same tools developed in Section 1.1.1. Hence, the definition of the NC Wigner function may be attained by the scheme [15]:

$$\hat{G}(\tilde{z}) \xrightarrow{\hat{D}} \hat{G}(z) \xrightarrow{W} g(z) \xrightarrow{D^{-1}} g(\tilde{z}). \tag{1.50}$$

As the Darboux map is linear, no order ambiguity arises when applying the transformation, thus D and $\hat{D}$ in the above equation have the same functional form. This means that one is able to define the generalized WW transform by:

$$W^{NC}[\hat{G}] = \left(D^{-1} \circ W \circ D\right)[\hat{G}]. \tag{1.51}$$

The above definition of the NC Wigner-Weyl transform has several features which are desirable. These can be listed as:

(a) $W^{NC}[\hat{1}] = 1$,

(b) $W^{NC}[\hat{Q}] = q$,

(c) $W^{NC}[\hat{K}] = k$.

Thus, the identity in the space of operators corresponds to the identity in the space of functions, and the position and momentum operators are mapped to the usual position and momentum coordinates in the phase-space. Furthermore, the definition of W^{NC} is compatible with the deformed $\star$-product, Eq. (1.44), in the same sense as in Eq. (1.15) for the standard product, i.e.,

$$W^{NC}[\hat{A}\hat{B}] = W^{NC}[\hat{A}] \star_{NC} W^{NC}[\hat{B}] = \tilde{a} \star \tilde{b}, \tag{1.52}$$

where $\tilde{a} = \tilde{a}(q,k)$ represents the Weyl transform of $\hat{A}$ and the $\sim$ denotes the distinction between this function and the commutative one, marking this as one on the NC phase-space.

Given that definition Eq. (1.51) explicitly depends on the map D one might be concerned about the results obtained via W^{NC} also being dependent on this map. Yet, despite the explicit use of the map, if it is linear, all measurable quantities are independent of it [15]. Since the change in the commutation relations for NCQM is global, it is always possible to find a linear Darboux map. Thus, the obtained results are a consequence of the extra commutation relations on the algebra and not of the particular details of the linear map used to make calculations. The independence of this definition on the map is a direct consequence of its linearity. To see this, one can look at the Darboux map in two different ways. First, it is a coordinate transformation on phase-space, such that in the new coordinates the symplectic form takes the standard form. Second, it can be seen as a map from a set of operators, $(\hat{Q}, \hat{K})$, to another set, $(\hat{X}, \hat{P})$, which obeys standard commutation relations. These views are equivalent since the commutation relations are determined by the symplectic form, as seen in Eq. (1.43). Given this, and the fact that the Darboux map is linear, there are no ordering ambiguities in transforming operators into functions. Consequently, $\hat{D}$ and D have the same functional form and so $\hat{D}^{-1}$ undoes the effect of D, as the ordering of the operators before applying W is the same as the ordering of functions after it has been applied. A more rigorous proof of this statement can be found in [insert Catarina ref], where it is shown that the objects obtained from different maps are related through a unitary transformation, thus making observables invariant under map choice.

For the description of the state of a quantum system, one can use Eq. (1.51) to compute the noncommutative Wigner function for a given state, by applying the NC WW transform to the density matrix of the system, which them yields $f^{NC}(\tilde{z})$. However, this definition of the noncommutative Wigner function is not normalized in the usual sense. Instead it obeys [15]:

$$\int \frac{1}{\sqrt{\det\Omega}} f^{NC}(\tilde{z}) \mathrm{d}^n \tilde{z} = 1, \tag{1.53}$$

with the $\det\Omega$ factor coming from the coordinate change in the WW transform integral. As our map D is linear, Ω has constant entries which implies that $\det\Omega$ is also constant. Hence, one may re-define the noncommutative Wigner function as:

$$f^{NC}(\tilde{z}) \rightarrow \frac{1}{\sqrt{\det\Omega}} f^{NC}(\tilde{z}), \tag{1.54}$$

as this only changes the original function by a numeric factor. Also, this definition still allows for the recovery of $f(z)$ in the limit η, $\theta = 0$ as $\lim \det\Omega = 1$. Moreover, it allows us to write the expectation value of an observable in NCQM as:

$$\langle g \rangle_{NC} = \int f^{NC}(\tilde{z}) \tilde{g}(\tilde{z}) \mathrm{d}^n \tilde{z}. \tag{1.55}$$

The only missing ingredient in this treatment of NCQM is the differential equations that describe the states of the system as well as their evolution. These are, it turns, remarkably similar to the ones that describe QM in the phase-space, as a result of the modifications to the $\star$-product and the Wigner-Weyl transform. The NC Moyal equation is given by:

$$\frac{\partial f^{NC}(\tilde{z})}{\partial t} = \{\!\{ \tilde{H}(\tilde{z}), f^{NC}(\tilde{z}) \}\!\}_{NC}, \tag{1.56}$$

where the Moyal brackets on the left hand side of the equation above are defined with the modified $\star$-product, Eq. (1.44). In turn, the $\star$-gen equation for the NC phase-space is given by:

$$\tilde{H}(\tilde{z}) \star_{NC} f^{NC}(\tilde{z}) = E f^{NC}(\tilde{z}). \tag{1.57}$$

One should notice that both Eq. (1.56) and Eq. (1.57) are invariant under the redefinition of the NC WF, Eq. (1.54). This is to be expected, as a re-definition of this function should not affect the physical results that one obtains from the theory.

The above treatment of the PS formalism of QM that deals with the definition of the important tools when taking a coordinate change encompasses more than its use for PSNCQM. In fact, this treatment allows for a formulation that is invariant under general coordinate changes, in the sense that the equations of motion are the same regardless of the coordinate system, as long as the $\star$-product and WW transform are redefined through Eqs. (1.44) and (1.51), respectively.

As a final remark, it should be pointed out that the NC version of this formalism is not necessarily defined in terms of the commutative one. As a matter of fact, one could start with the deformed $\star$- product and write Eqs. (1.56) and (1.57) to solve systems in NCQM and then use the inverse Darboux map D^{-1} to define the standard QM in terms of the deformed one. This would be the same to what has been done here in essence, since with this treatment the EoM are coordinate invariant. The advantage of the approach presented here is the fact that all the results were already known for the HW algebra and so the formulation in terms of these results allows for a direct comparison to assess the consequences of the introduction of the two NC fundamental constants θ and η. Hence, one can regard NCQM as a deformation in the Moyal $\star$-product of the theory, with unchanged equations of motion, whose solutions are now functions of θ and η.

1.1.5 Moyal $\star$-product

The Moyal $\star$-product for the commutative and NC theories can be written as:

$$a(\chi) \star_\Lambda b(\chi) = a(\chi)e^{\frac{i}{2}\overleftarrow{\partial}_{\chi_r}\Lambda_{rs}\overrightarrow{\partial}_{\chi_s}}b(\chi) \, ,\tag{1.58}$$

where, for $\star$ and $\star_{NC}$, the variable χ stands for $\chi = z$ in the $2n$-dimensional phase-space, and the symplectic matrices in these cases read [2]

$$\Lambda = \hbar\Omega, \quad \text{if } \star_\Lambda = \star_{NC} \qquad , \qquad \Lambda = \hbar J, \quad \text{if } \star_\Lambda = \star \, ,\tag{1.60}$$

with

$$J = \begin{pmatrix} 0 & I_{n\times n} \\ -I_{n\times n} & 0 \end{pmatrix} \quad \text{and} \quad \Omega = \begin{pmatrix} \frac{1}{\hbar}\Theta & I_{n\times n} \\ -I_{n\times n} & \frac{1}{\hbar}N \end{pmatrix}.\tag{1.61}$$

For Λ invertible, the $\star$-product of the form Eq. (1.58) admits a kernel representation:

$$a(\chi) \star_\Lambda b(\chi) = \frac{1}{\pi^n|\det\Lambda|}\int d\chi'\int d\chi'' \; a(\chi')b(\chi'')e^{[2i(\chi-\chi')^T\Lambda^{-1}(\chi''-\chi)]},\tag{1.62}$$

for[3] $a,b \in \mathcal{A}(\mathbb{R}^n)$.

---

[2] For $\star_\theta$ and $\star_\eta$, the variable χ stands for $\chi = \{x_i\}$ or $\chi = \{p_i\}$ $(r,s = 1,\cdots,n)$, and the symplectic matrices read $\Lambda = \Theta$, if $\star_\Lambda = \star_\theta$ and $\Lambda = N$, if $\star_\Lambda = \star_\eta$, respectively. According to Ref. [15], a $\star$-product of the form Eq. (1.58) acting on the space of polynomials on phase-space ($\star_\Lambda = \star$ or $\star_\Lambda = \star_\hbar$), configuration space ($\star_\Lambda = \star_\theta$) or momentum space ($\star_\Lambda = \star_\eta$), can be represented as a Bopp shift,

$$a(\chi) \star_\Lambda b(\chi) = a\left(\chi + \frac{i}{2}\Lambda\overrightarrow{\partial}_\chi\right)b(\chi) = a(\chi)b\left(\chi - \frac{i}{2}\Lambda\overleftarrow{\partial}_\chi\right).\tag{1.59}$$

[3] Where n stands for $2n$ or n depending on whether the $\star$-product is $\star$, $\star_\hbar$, or $\star_\theta$, $\star_\eta$.

As an example, one takes the particular deformation of PS NCQM. To find the correspondence between ordinary and NCQM from the above results, one performs the SW transformation,

$$\mathbf{z} \longrightarrow \tilde{\mathbf{z}} = \mathbf{S}\mathbf{z} \ , \ \ \alpha(\mathbf{z}) \longrightarrow \alpha'(\tilde{\mathbf{z}}) = \alpha(\mathbf{z}(\tilde{\mathbf{z}})) \ , \ \ \beta(\mathbf{z}) \longrightarrow \beta'(\tilde{\mathbf{z}}) = \beta(\mathbf{z}(\tilde{\mathbf{z}})), \tag{1.63}$$

with $S_{km} = \frac{\partial \tilde{z}_k}{\partial z_m}$, which reproduces the calculations for the corresponding commutative variables, $\alpha(\mathbf{z})$ and $\beta(\mathbf{z})$. The kernel representation of the Moyal $\star$-product is given by

$$\alpha(\mathbf{z}) \star_\hbar \beta(\mathbf{z}) = \frac{1}{(\pi\hbar)^{2n}} \int d\mathbf{z}' \int d\mathbf{z}'' \ \alpha(\mathbf{z}')\beta(\mathbf{z}'')e^{[-\frac{2i}{\hbar}(\mathbf{z}-\mathbf{z}')^T J(\mathbf{z}''-\mathbf{z})]}. \tag{1.64}$$

Given that the symplectic matrix transforms as $\mathbf{\Omega} = \mathbf{S}\mathbf{J}\mathbf{S}^T$, and, of course, $\det \mathbf{S} = \sqrt{\det \mathbf{\Omega}}$, under the SW map, Eq. (1.64) transforms according to [15]:

$$\alpha(\mathbf{z}) \star_\hbar \beta(\mathbf{z})\Big|_{\mathbf{z}=\mathbf{z}(\tilde{\mathbf{z}})} = \frac{1}{(\pi\hbar)^{2n}} \int d\tilde{\mathbf{z}}' \int d\tilde{\mathbf{z}}'' (\det \mathbf{S})^{-2}\alpha'(\tilde{\mathbf{z}}')\beta'(\tilde{\mathbf{z}}'')e^{[-\frac{2i}{\hbar}(\tilde{\mathbf{z}}-\tilde{\mathbf{z}}')^T (\mathbf{S}^{-1})^T J\mathbf{S}^{-1}(\tilde{\mathbf{z}}''-\tilde{\mathbf{z}})]},$$
$$\tag{1.65}$$

and, therefore,

$$\alpha'(\tilde{\mathbf{z}}) \star \beta'(\tilde{\mathbf{z}}) = \frac{1}{(\pi\hbar)^{2n}|\det \mathbf{\Omega}|} \int d\tilde{\mathbf{z}}' \int d\tilde{\mathbf{z}}'' \ \alpha'(\tilde{\mathbf{z}}')\beta'(\tilde{\mathbf{z}}'')e^{[\frac{2i}{\hbar}(\tilde{\mathbf{z}}-\tilde{\mathbf{z}}')^T \mathbf{\Omega}^{-1}(\tilde{\mathbf{z}}''-\tilde{\mathbf{z}})]}, \tag{1.66}$$

through which, for $a(\tilde{\mathbf{z}}) \equiv \alpha'(\tilde{\mathbf{z}})$ and $b(\tilde{\mathbf{z}}) \equiv \beta'(\tilde{\mathbf{z}})$, one recovers Eq. (1.62). A useful result that arises from the kernel representation of the Moyal $\star$-product, which as seen can be written both for regular QM and for NCQM is stated in the following theorem [15]:

Theorem: *For a Moyal $\star$-product of the form Eq. (1.62), one has*

$$\int d\chi \ A(\chi) \star_\Lambda B(\chi) = \int d\chi \ A(\chi)B(\chi), \tag{1.67}$$

similar to the WWGM formulation of ordinary QM.

From this theorem, whose proof can be found in Ref. [15], it can be seen that, since general $\star$-product possesses the same features regardless of the algebra of the phase-space coordinates, one can use the generalized Weyl-Wigner map to define the NC Wigner function.

1.2 The ADM formalism

1.2.1 3+1 decomposition

In order to proceed to the quantization of a certain manifold, it is useful to formulate the action of the theory using the Hamiltonian, rather than the Lagrangian. In order to do that, the manifold is foliated into a set of spacelike three dimensional surfaces, as:

$$\mathcal{M} = \mathbf{R} \times \mathcal{N}. \tag{1.68}$$

Here $\mathcal{M}$ is the four-dimensional manifold and $\mathcal{N}$ are the three-dimensional hypersurfaces. Let us now consider a set of coordinates x^α on $\mathcal{M}$ that cover the entire manifold. It is also useful to introduce a scalar field $t = t(x^\alpha)$, a single-valued function of x^α, for which the condition $t = constant$ describes a family of three-dimensional hypersurfaces with no intersections. Each of these must be labeled by the value of the scalar field on it, i.e., Σ_t. Let us also demand the normal vector to Σ_t, $n^\alpha \propto \partial^\alpha t$, to be oriented in the increasing direction of $t(x^\alpha)$. Now, for each surface Σ_t, let us introduce coordinates y^a, so that one is able to use the coordinate system (t, y^a). In a sense, the spacetime is now foliated into spacelike hypersurfaces, giving "time" a special place in the coordinate system, so this setup is more convenient to write the GR Hamiltonian.

We must now consider a congruence γ that intersects each hypersurface only once. Since each surface has a fixed value of t, this is a good parameter for the curves on the congruence. Selecting a specific curve on the congruence, γ_P, that intersects the hypersurface Σ_t at point P, it is clear that it forms a map from points on Σ_t to points P_i on Σ_{t_i}. Thus, in order to relate the coordinates y^a on each hypersurface, we impose that the points which are intersected by the same curve have the same y coordinate, i.e., $y^a(P) = y^a(P_i)$, $\forall\, P_i$ in γ_P.

Given the previous construction, the vector tangent to the congruence, t^α, is given by:

$$t^\alpha = \left(\frac{\partial x^\alpha}{\partial t} \right), \tag{1.69}$$

which can be defined since there is a relation between coordinate systems, $x^\alpha(t, y^a)$. Also, we can define vector tangent to Σ_t, which can be written as:

$$e_a^\alpha = \left(\frac{\partial x^\alpha}{\partial y^a} \right), \tag{1.70}$$

where "a" labels the vector and "α" labels the components of each vector. We must note that, with these definitions, the Lie derivative yields:

$$(\mathcal{L}_t e_a)^\alpha = g^{\alpha\sigma} \left(t^\beta \partial_\beta (e_a)_\sigma + (e_a)_\beta \partial_\sigma t^\beta \right) = \delta_t^\beta \partial_\beta (\delta_a^\alpha) + g^{\alpha\sigma} (e_a)_\beta \partial_\sigma (\delta_t^\beta) = 0, \tag{1.71}$$

where in the second step we used the fact that, in (t, y^a) coordinates, $t^\alpha = \delta_t^\alpha$ and $e_a^\alpha = \delta_a^\alpha$. Thus, since the Lie derivative relative to t^α is null, the vectors e_a^α are not transported along this vector field, being in fact tangent to Σ_t. It must be noted at this point that the vector tangent to γ is not necessarily normal to Σ_t, since the curves do not cross the hypersurfaces orthogonally. The vector normal to this surfaces is $\partial_\alpha t$, which, due to the open choice of the "time" scalar field, may happen not to be normalized. Thus, we introduce the unit normal to the surface, writen as:

$$n_\alpha = -N\partial_\alpha t. \tag{1.72}$$

The scalar field N is called the *lapse* function and it guarantees the correct normalization of the normal vector at any point. The minus sign ensures that the normal vector is future oriented. Clearly, since this is orthogonal to Σ_t, condition

$$n_\alpha e_a^\alpha = -N\partial_\alpha t\, \delta_a^\alpha = -N\partial_a t = 0 \tag{1.73}$$

holds, which is shown using (t, y^a) coordinates. One now have an orthonormal basis of vectors that generate the tangent space at each point, $\{n^\alpha, e_a^\alpha\}$. Hence, it is possible to write t^α in this basis, which becomes:

$$t^\alpha = Nn^\alpha + N^a e_a^\alpha. \tag{1.74}$$

The three-vector N^a gives the decomposition of t^α tangent to Σ_t and is called the *shift* vector. Given this decomposition, and using Eqs. (1.72) and (1.73) we can write:

$$\begin{aligned}
t^\alpha \partial_\alpha t &= \left(Nn^\alpha + N^a e_a^\alpha\right)\partial_\alpha t = Nn^\alpha \partial_\alpha t + N^a e_a^\alpha \partial_\alpha t = \\
&= -N^2 g^{\alpha\beta}\partial_\alpha t\partial_\beta t = -N^2 g^{tt},
\end{aligned} \tag{1.75}$$

where in the last step (t, y^a) coordinates were used. On the other hand, it is known that $t^\alpha \partial_\alpha t = 1$, so that we get:

$$g^{tt} = -\frac{1}{N^2} \tag{1.76}$$

We are now in position to write the metric of the manifold $\mathcal{M}$ in this foliated set of coordinates. It is clear that an infinitesimal displacement along the congruence γ can be written as a sum of the tangencial displacement plus the displacement parallel to Σ_t, that is:

$$dx^\alpha = t^\alpha\, dt + e_a^\alpha\, dy^a = Nn^\alpha dt + \left(N^a dt + dy^a\right)e_a^\alpha. \tag{1.77}$$

It follows that:

$$ds^2 = g_{\alpha\beta}dx^\alpha dx^\beta = -N^2 dt^2 + h_{ab}\left(N^a dt + dy^a\right)\left(N^b dt + dy^b\right), \tag{1.78}$$

where it was defined $h_{ab} := g_{\alpha\beta}e_a^\alpha e_b^\beta$, the induced metric in the hypersurface Σ_t. This form of the metric will be of much use in further sections of this work, when quantization of specific manifolds is performed. Also, in writing the action for GR one needs to deal with $\sqrt{-g}$.

Using the notation $g := \det(g_{\alpha\beta})$, we can write:

$$g = g^{tt} \text{ cofactor} \left(g^{tt} \right), \tag{1.79}$$

since $g^{t\alpha} = 0$ for $\alpha \neq t$. Hence, as cofactor $\left(g^{tt} \right) = \det(h_{ab}) := h$ we can write:

$$g = -N^2 h \Leftrightarrow \sqrt{-g} = N\sqrt{h}. \tag{1.80}$$

Equations (1.74), (1.78) and (1.80) are the main result of $3 + 1$ decomposition, which allow for writing the Hamiltonian for GR.

1.2.1.1 Boundary foliation

Before proceeding to the Hamiltonian description of GR, it is useful to consider the boundary of spacetime, since there is a boundary term in the Einstein-Hilbert action, the Gibbons-Hawking-York boundary term, that must be taken into account when the underlying spacetime has a boundary. This is given by:

$$S_{\text{boundary}} = \frac{1}{8\pi} \oint_{\partial\mathcal{V}} \epsilon K |h|^{1/2} \mathrm{d}^3 y, \tag{1.81}$$

where h is the determinant of the induced metric on Σ_t, $\epsilon = n^\alpha n_\alpha$ and K is the trace of the extrinsic curvature of Σ_t, defined by:

$$K_{ab} := n_{\alpha;\beta}\, e_a^\alpha\, e_b^\beta. \tag{1.82}$$

This quantity is also called the second fundamental form in the language of differential geometry. The semi-colon denotes a covariant derivative with the connection being given by the Christofell symbols.

In order to write this term in a way that is more suitable for the Hamiltonian formulation, one first needs to define a two dimensional surface S_t which is the boundary of Σ_t, that is, $S_t := \partial\Sigma_t$. Then, let the coordinates in S_t be θ^A, defined by some parametric relation $y^a(\theta^A)$, and define r_a to be the unit normal to this surface. An associated four-vector may be constructed as:

$$r^\alpha = r^a\, e_a^\alpha, \tag{1.83}$$

in such a way that in (t, y^a) coordinates the t component vanishes, so that it is a vector on Σ_t. For this vector:

$$r^\alpha r_\alpha = g_{\alpha\beta}\, r^a \delta_a^\alpha r^b \delta_b^\beta = r^a r_a = 1, \tag{1.84}$$

where in the second step (t, y^a) coordinates were used. We also have that:

$$r^\alpha n_\alpha = r^a\, e^\alpha_a n_\alpha = 0,\tag{1.85}$$

so that these vectors are orthogonal. A basis for S_t can be constructed considering the projections of the y^a coordinates in the θ^A coordinates, which can be written as:

$$e^a_A := \frac{\partial y^a}{\partial \theta^A},\tag{1.86}$$

for which:

$$r_a\, e^a_A = 0.\tag{1.87}$$

For this to be useful in the four dimensional manifold, we define an associated four-vector:

$$e^\alpha_A := e^\alpha_a\, e^a_A = \frac{\partial x^\alpha}{\partial \theta^A}.\tag{1.88}$$

This vector is the four dimensional extension of e^a_A and obeys $r_\alpha\, e^\alpha_A = r_a\, e^a_A = 0$ because by definition the vectors in the last step are orthogonal. We now have a new set of coordinates, in which the y^a coordinates get split, given by (t, r, θ^A). In these, r is the coordinate orthogonal to S_t and θ^A are the coordinates of that surface. In this coordinates, the induced metric in S_t is given by:

$$ds^2 = \sigma_{AB}d\theta^A d\theta^B.\tag{1.89}$$

Using (t, r, θ^A) coordinates, we can see that $\sigma_{AB} = h_{ab}e^a_A e^b_B = g_{\alpha\beta}e^\alpha_A e^\beta_B = g_{AB}$, so that the induced metric is composed by the components of the original metric that correspond to the coordinates θ^A. In the last step we used the fact that $e^\alpha_A = \delta^\alpha_A$ in (t, r, θ^A). An additional structure can be computed for S_t, considering it is embedded in Σ_t: the extrinsic curvature. This is defined in a similar way to the one of Σ_t embedded in $\mathcal{M}$, being given by:

$$k_{AB} := r_{\alpha;\beta}\, e^\alpha_A e^\beta_B.\tag{1.90}$$

The last issue regarding the boundary term is similar to one already treated in Subsection 1.2.1 and it concerns the relationship between coordinates at different S_t. To address this, let us define a three-surface $\mathcal{B}$ by:

$$\mathcal{B} = \bigcup_t S_t,\tag{1.91}$$

and also consider a congruence of curves, η that cover $\mathcal{B}$ and that intersect the surfaces S_t orthogonally, that is, the tangent vector to a certain curve η_P of the congruence that contains point P is at any point the vector normal to the surface S_t, n^α, $\forall\, t$. The surface $\mathcal{B}$ can be regarded as the spatial boundary at any time t. Let us remember that the coordinates in S_t

are (θ_A), hence the coordinates in $\mathcal{B}$ are (t, θ_A). Then, in order to relate points in different S_t, we demand that points in the same curve of the congruence all have the same θ_A coordinates. This construction ensures that the basis vector of each S_t and n^α are everywhere orthogonal. This property will be useful in describing the metric of this surface but first let us set another coordinate system in $\mathcal{B}$: (z_i). The tangent vectors to it are thus given by:

$$e_i^\alpha := \frac{\partial x^\alpha}{\partial z_i}, \tag{1.92}$$

and the induced metric in $\mathcal{B}$ is:

$$\gamma_{ij} = g_{\alpha\beta} e_i^\alpha e_j^\beta. \tag{1.93}$$

Choosing now $(z_i) = (t, \theta_A)$ leads to:

$$\mathrm{d}x^\alpha = Nn^\alpha \mathrm{d}t + e_A^\alpha \mathrm{d}\theta^A \tag{1.94}$$

$$\Rightarrow \mathrm{d}s_\mathcal{B}^2 = g_{\alpha\beta}\mathrm{d}x^\alpha \mathrm{d}x^\beta = -N^2\mathrm{d}t^2 + \sigma_{AB}\mathrm{d}\theta^A\mathrm{d}\theta^B, \tag{1.95}$$

where the relation $\sigma_{AB} = g_{\alpha\beta} e_A^\alpha e_B^\beta$ and the orthogonality of n^α and e_A^α were used. Since we can also write:

$$\mathrm{d}s_\mathcal{B}^2 = \gamma_{ij}\mathrm{d}z^i\mathrm{d}z^j, \tag{1.96}$$

the determinants of the metric in S_t and the induced metric in $\mathcal{B}$ are related through:

$$\sqrt{-\gamma} = N\sqrt{\sigma}. \tag{1.97}$$

Finally, using the coordinates we have set in $\mathcal{B}$, we can define the extrinsic curvature for the surface in a way very similar to k_{AB} as:

$$\mathcal{K}_{ij} := r_{\alpha;\beta}\, e_i^\alpha\, e_j^\beta. \tag{1.98}$$

The coordinates defined in this Subsection will be of major importance in the following treatment of the GR action and in the Hamiltonian formulation.

1.2.2 Action and Hamiltonian

Now that all the formalism is in place, it is time to proceed to the Hamiltonian formulation. In order to achieve this, we must first make the time dependence of the action explicit. We therefore start with the Einstein-Hilbert action, which is written as:

$$S = \frac{1}{16\pi}\int_\mathcal{V} \sqrt{-g}R\mathrm{d}^4x + \frac{1}{8\pi}\oint_{\partial\mathcal{V}} \epsilon K|h|^{1/2}\mathrm{d}^3y, \tag{1.99}$$

where K is the trace of the extrinsic curvature of the boundary and which we assume to be:

$$\partial\mathcal{V} = \Sigma_{t_2} \cup \Sigma_{t_1} \cup \mathcal{B}, \tag{1.100}$$

with $t_2 > t_1$. With such decomposition of the boundary, and using Eqs. (1.90) and (1.98), we can write the boundary action as:

$$S_{boundary} = \frac{1}{8\pi} \oint_{\Sigma_{t_2}} \epsilon K |h|^{1/2} \mathrm{d}^3 y - \frac{1}{8\pi} \oint_{\Sigma_{t_1}} \epsilon K |h|^{1/2} \mathrm{d}^3 y + \frac{1}{8\pi} \oint_{\mathcal{B}} \epsilon K | - \gamma|^{1/2} \mathrm{d}^3 z. \tag{1.101}$$

In the last equation we used $K = K_{ab} h^{ab}$ and $K = K_{ij} \gamma^{ij}$. Regarding the bulk term, it can be shown [32] that:

$$R = {}^3R + K^{ab} K_{ab} - K^2 - 2 \left(n^\alpha_{;\beta} n^\beta - n^\alpha n^\beta_{;\beta} \right)_{;\alpha}. \tag{1.102}$$

The last term in the previous equation generates a boundary term by application of the divergence theorem, which can be once more split into the three boundary terms (see Eq. (1.100)). Moreover, using Eq. (1.80), we can write the measure in the bulk as $\sqrt{-g}\,\mathrm{d}^4 x = N\sqrt{h}\,\mathrm{d}t\,\mathrm{d}^3 y$, thus splitting the integration in time from the one in spatial degrees of freedom. This is the first step into giving time a "special" role. Using these relations, we can write the first term in Eq. (1.99) as:

$$\int_{t_1}^{t_2} \mathrm{d}t \int_{\Sigma_t} \left({}^3R + K^{ab} K_{ab} - K^2 \right) N\sqrt{h}\,\mathrm{d}^3 y - 2 \oint_{\Sigma_{t_2} \cup \Sigma_{t_1} \cup \mathcal{B}} \left(n^\alpha_{;\beta} n^\beta - n^\alpha n^\beta_{;\beta} \right) \epsilon\, n_\alpha \sqrt{h}\,\mathrm{d}^3 y, \tag{1.103}$$

where we used the fact that the Σ_t surface vector is given by $\mathrm{d}\Sigma_\alpha = n_\alpha \sqrt{h}\,\mathrm{d}^3 y$ and ϵ accounts for the direction of the normal vector. Considering that on Σ_{t_1} the normal vector must point outwards, $\epsilon = -1$ and thus the integration of the last term in Eq. (1.103) over this surface leads to:

$$2 \oint_{\Sigma_{t_1}} \left(n^\alpha_{;\beta} n^\beta - n^\alpha n^\beta_{;\beta} \right) n_\alpha \sqrt{h}\,\mathrm{d}^3 y = 2 \oint_{\Sigma_{t_1}} \left(n^\alpha_{;\beta} n^\beta n_\alpha + n^\beta_{;\beta} \right) \sqrt{h}\,\mathrm{d}^3 y. \tag{1.104}$$

On the other hand:

$$n^\alpha_{;\beta} n^\beta n_\alpha = (n^\alpha n_\alpha)_{;\beta}\, n^\beta - n^\alpha n_{\alpha;\beta} n^\beta \Rightarrow n^\alpha_{;\beta} n^\beta n_\alpha = 0, \tag{1.105}$$

and also:

$$K = h^{ab} K_{ab} = h^{ab} n_{\alpha;\beta} e^\alpha_a e^\beta_b = h^{ab} n_{\alpha;\beta} \delta^\alpha_a \delta^\beta_b = n^\beta_{;\beta}, \tag{1.106}$$

so that Eq. (1.104) becomes:

$$2 \oint_{\Sigma_{t_1}} K \sqrt{h}\,\mathrm{d}^3 y, \tag{1.107}$$

which cancels out the second term in Eq. (1.101). A similar process can be carried for the integration over Σ_{t_2} which cancels the appropriate term in Eq. (1.101). Hence, the remaining boundary term comes from the integration in $\mathcal{B}$. On this surface the measure of integration is $dS_\alpha = r_\alpha \sqrt{-\gamma}\, d^3z$ and since $n^\alpha r_\alpha = 0$ by construction we have:

$$\left(n^\alpha_{;\beta} n^\beta - n^\alpha n^\beta_{;\beta} \right) r_\alpha = n^\alpha_{;\beta} n^\beta r_\alpha = -r_{\alpha;\beta} n^\alpha n^\beta, \tag{1.108}$$

so that Eq. (1.99) can be written as:

$$S = \frac{1}{16\pi} \left[\int_{t_1}^{t_2} dt \int_{\Sigma_t} \left({}^3R + K^{ab} K_{ab} - K^2 \right) N\sqrt{h}\, d^3y + 2 \int_{\mathcal{B}} \left(\mathcal{K} + r_{\alpha;\beta} n^\alpha n^\beta \right) \sqrt{-\gamma}\, d^3z \right]. \tag{1.109}$$

We can now use the coordinates (t, θ_A) set in $\mathcal{B}$ to write $\sqrt{-\gamma}\, d^3z = N\sqrt{\sigma}\, dt\, d^2\theta$. Furthermore, using Eq. (1.98), (1.90) and the orthogonality relations for n^α and r^α it is easy to obtain the following relation:

$$\mathcal{K} + r_{\alpha;\beta} n^\alpha n^\beta = \sigma^{AB} k_{AB} = k. \tag{1.110}$$

Therefore, the action can be cast in a form where time has a separate role, namely:

$$S = \frac{1}{16\pi} \left\{ \int_{t_1}^{t_2} dt \left[\int_{\Sigma_t} \left({}^3R + K^{ab} K_{ab} - K^2 \right) N\sqrt{h}\, d^3y + 2 \oint_{S_t} k N\sqrt{\sigma}\, d^2\theta \right] \right\}, \tag{1.111}$$

apart from boundary constants to prevent divergence. This equation separates time from the spatial degrees of freedom, allowing for the definition of an Hamiltonian for GR. It must be noted that the degrees of freedom for which a variation must be taken to find the equations of motion are at this point N, N_a and h_{ab}, since these quantities suffice to define the metric and their variation is equivalent to the variation of $g_{\alpha\beta}$ in Eq. (1.99).

In order to conclude the Hamiltonian treatment of GR it is useful to introduce the conjugate momenta of the independent variables mentioned above. Regarding the lapse and shift functions, respectively, it can be written:

$$p_N := \frac{\partial \mathcal{L}}{\partial \dot{N}}, \qquad p_a := \frac{\partial \mathcal{L}}{\partial \dot{N}^a}, \tag{1.112}$$

both of which vanish since the Lagrangian (see Eq. (1.111)) does not depend on these variables. Defining the time derivative of the induced metric as $\dot{h}_{ab} := \pounds_t h_{ab}$, the conjugate momentum to h_{ab} is given by:

$$p^{ab} := \frac{\partial \mathcal{L}}{\partial \dot{h}_{ab}} = \frac{\sqrt{h}}{16\pi} \left(K^{ab} - K h^{ab} \right) = \frac{1}{16\pi} G^{abcd} K_{cd}, \tag{1.113}$$

wherein it was used that:

$$G^{abcd} := \frac{\sqrt{h}}{2} \left(h^{ac} h^{bd} + h^{ad} h^{bc} - 2h^{ab} h^{cd} \right). \tag{1.114}$$

This tensor is called the *DeWitt metric*. Also using the definition in Eq. (1.114) in Eq. (1.111) it is possible to write the bulk action term as:

$$\left({}^3R + K^{ab}K_{ab} - K^2\right) N\sqrt{h} = \left({}^3R\sqrt{h} + G^{abcd}K_{ab}K_{cd}\right) N. \tag{1.115}$$

Then, using the identity [33]:

$$\dot{h}_{ab} = 2NK_{ab} + N_{a|b} + N_{b|a}, \tag{1.116}$$

were the bar denotes the covariant derivative with respect to h_{ab} and Eq. (1.113) the derivative of this metric becomes:

$$\dot{h}_{ab} = \frac{32\pi N}{\sqrt{h}}\left(p_{ab} - \frac{1}{2}ph_{ab}\right) + N_{a|b} + N_{b|a}. \tag{1.117}$$

If we are treating unbounded manifolds we can neglect the boundary term of Eq. (1.111) and focus on the bulk term. Thus it is then straightforward to define the GR Hamiltonian in the usual manner as:

$$\mathcal{H} := p^{ab}\dot{h}_{ab} - \mathcal{L}, \tag{1.118}$$

which upon recalling Eqs. (1.111), (1.113) and (1.117) becomes:

$$\mathcal{H} = 16\pi N G_{abcd}p^{ab}p^{cd} - N\frac{\sqrt{h}\,{}^3R}{16\pi} - 2N_b p^{ab}_{|a}. \tag{1.119}$$

The bulk action is then written as:

$$S = \frac{1}{16\pi}\int_{t_1}^{t_2} dt \int d^3y \left(p^{ab}\dot{h}_{ab} - N\mathcal{H}_N - N^a\mathcal{H}_a\right), \tag{1.120}$$

where:

$$\mathcal{H}_N = 16\pi G_{abcd}p^{ab}p^{cd} - \frac{\sqrt{h}\,{}^3R}{16\pi}, \qquad \mathcal{H}_a = -2D_b p^b_a. \tag{1.121}$$

Upon variation of the action with respect to the lapse and shift functions, four non-dynamical equations are obtained:

$$\mathcal{H}_N = 0, \qquad \mathcal{H}_a = 0. \tag{1.122}$$

These are known as the Hamiltonian constraints and are the necessary equations to quantize GR. The dynamical equations obtained via variation with respect to h_{ab} and p^{ab} are not important for the quantization procedure.

1.2.3 Wheeler-DeWitt equation

Following the steps of the previous section it is rather straightforward to obtain the quantized equation. The main constraint to the theory is given in Eq. (1.122). In order to obtain a quantized theory we must simply state that all functions become operators acting on an yet undefined Hilbert space. Thus the Wheeler-DeWitt (WDW) equation reads:

$$\hat{\mathcal{H}}_N \Psi\left[h_{ab}\right] = 0, \tag{1.123}$$

where $\Psi\left[h_{ab}\right]$ is a functional of the induced metric. This functional must not depend on p^{ab} since we must impose canonical commutation relations on h_{ab} and p^{ab}. This happens because the Poisson brackets of these quantities can be shown to be:

$$\{h_{ab}(x), p^{cd}(y)\} = \delta^c_{(a}\delta^d_{b)}\delta(x,y), \tag{1.124}$$

hence the commutation relation between operators $\hat{h}_{ab}$ and $\hat{p}^{cd}$ must be defined to be:

$$\left[\hat{h}_{ab}(x), \hat{p}^{cd}(y)\right] = \mathrm{i}\delta^c_{(a}\delta^d_{b)}\delta(x,y), \tag{1.125}$$

and so Ψ must only be a functional of one of these variables. Moreover, in this "metric representation" we can write the metric and momentum operators acting on Ψ as:

$$\hat{h}_{ab}\Psi \to h_{ab}\Psi, \qquad \hat{p}^{ab}\Psi \to -\mathrm{i}\frac{\delta}{\delta h_{ab}}\Psi, \tag{1.126}$$

where δ stands for the functional derivative. In the particular case of minisuperspace models it reduces to a standard partial derivative. A solution to Eq. (1.123) leads to a wavefunction of the manifold in study. One must also notice that, in Eq. (1.123), the Hamiltonian is no longer the gererator of time evolution for the system. This property is known as *timelessness*. Additionally, the interpretation of the Hamiltonian and the state-functional, $\Psi[h_{ab}]$, are distinct from their interpretation in non-relativistic quantum mechanics. Here, $\Psi[h_{ab}]$ contains the information about the possible matter and geometry information for the universe (i.e., all of spacetime). The Hamiltonian then acts as a contraint on the functional Ψ, stating that the physical states (i.e., configurations for the Universe) are those which vanish in the sense of Eq. (1.123). Thus, it acts as a constraint on the Hilbert space of Universe configurations. When performing the quantization of the constraint in the first equation of Eq. (1.122), which leads to the WdW equation, some caution is advised due to operator ordering, since, unlike the real (or complex) function in Eq. (1.122), the operators in Eq. (1.123) do not commute, instead obeying to Eq. (1.125). This is a well known issue in the canonical quantization prodecure, since the quantum Hamiltonian is not uniquely defined by the classical limit, even if the quantum one is required to be Hermitian. This is an expectable problem to arise, since the quantization procedure being used is rather naive in this regard. Throughout this book, whenever the issue of the order ambiguity arises, one chooses the simplest ordering possible for the quantum version of the Hamiltonian compatible with the classical limit. When aplicable, this issue is discussed within the appropriate context.

Chapter 2

Quantum cloning and teleportation fidelity in the noncommutative phase-space

The manipulation of quantum states for the processing of information has been stirred up by an increasing number of theoretical tools which describe the prospects of quantum-enhanced systems. This includes several protocols for quantum factoring, quantum cloning and quantum teleportation, all of them involving multipartite quantum systems. In the context of these procedures, a series of limitations and assumptions have been established by the so-called no-cloning theorem, which ensures that no random state can be duplicated [34]. More specifically, the no-cloning theorem precludes the possibility of creating an auxiliary duplicate of a state during a quantum computation. Fortunately, the advent of quantum error correcting codes [35, 36] allows for circumventing the limitations of the no-cloning hypothesis of discrete states so to provide the setup for more general quantum computing protocols which involve, for instance, continuous variable framework [37]. Although originally designed for discrete variable quantum systems, the quantum platforms built from the continuous-spectrum described by the quadrature components of a light mode are presumably easier to handle than their discrete counterparts [38]. In fact, quantum teleportation [37] and quantum computation [39] protocols, as well as quantum cryptographic schemes [40] have been developed, all relying on continuous variables.

The aim of this Chapter is to investigate the foundations and the range of applicability of the no-cloning theorem in the framework of phase-space noncommutative (NC) quantum mechanics (QM). Our analysis comprises a covariant formulation of quantifiers of the fidelity of states and a continuous variable quantum teleportation protocol in terms of NC variables.

Our results clearly indicate that the no-cloning theorem can be generalized for continuous variables in the NC phase-space. In fact, it will be shown that cloning and teleportation probabilities obtained from NC Wigner functions can be quantified through a NC covariant expression for the teleportation fidelity that allows for quantifying the reproducibility of quantum cloned and teleported Wigner functions, which fits perfectly the results for ordinary QM of Gaussian states.

The outline of this Chapter is as follows. In Sec. 2.1, the formulation of the NC QM as discussed in Ref. [15] is briefly reviewed with some results specific for this Chapter. In particular, one will be concerned with the Wigner formulation of QM suitable for NCQM and the use of a generalized Seiberg-Witten (SW) map [13] to define the NC Wigner function and some of its properties which will be relevant in the following sections. The no-cloning theorem in the phase-space NC QM is obtained in Sec. 2.3. Sec. 2.4 addresses the covariant formulation of the teleportation fidelity in the NC phase-space. Since the cloning probabilities are also related to the entanglement fidelity, a relationship given in terms of NC Wigner functions is obtained and applied for quantifying the cloning probabilities of NC Gaussian states. As will be seen, in the transition from ordinary QM to NCQM, the entanglement fidelity exhibits the same covariant behavior of the Wigner functions, a fundamental feature in demonstrating the no-cloning theorem in the NC framework. Finally, in Sec. 2.5, a protocol for a teleportation process in phase-space is constructed so to provide an additional consistency test for reproducing quantum states in the NC framework.

2.1 NC QM in phase-space

The phase-space formalism of QM based on the Wigner function relies on the use of phase-space functions dual to quantum operators, via the Wigner-Weyl transform. For an operator, $\hat{A}$, the Wigner-Weyl transform, Eq. (1.12), maps quantum states into Wigner functions. The transform of the density matrix, $W[\hat{\rho}] := W(q, k)$, is a fundamental object of this formalism. However, since they are regular functions in the phase-space, i.e. functions of positions and momenta, q, and, k, the corresponding quantum behavior is encoded into the naturally defined $\star$-product, the Moyal product [30]. This QM formalism is equivalent and independent of the Dirac formalism, with the Wigner-Weyl transform acting as a dictionary between the two descriptions.

A particularly useful result in the computation of the expectation value of an operator is Eq. (1.11). From the above formula, and considering that $W[\hat{1}] = 1$, one obtains that, for normalized states (as seen in Eq. 1.13):

$$1 = \int W(q, k)\mathrm{d}^n q \, \mathrm{d}^n k. \tag{2.1}$$

Additionally, the overlap of two states, denoted here as P, can be written as:

$$P = |\langle \psi_a | \psi_b \rangle|^2 = \frac{1}{\hbar^2} \int W_a(q,k) W_b(q,k) \mathrm{d}^n q \, \mathrm{d}^n k,$$ (2.2)

with $W_i := W[\psi_i]$ denoting the Wigner-Weyl transform of the state $|\psi_a\rangle$. It is usual to denote the Wigner-Weyl transforms of operators by capital letters (i.e. $A(x,p)$) and the ones of states as $W[\psi]$. The Wigner function $W(x,p)$ can be regarded as a quasi-probability distribution on the phase-space, in the sense that it weighs the integration of the phase-space functions of all observables. Another relevant result arises when considering the state $\hat{A}|\psi\rangle$, for which the density matrix is given by

$$\hat{\rho}_A = \hat{A}|\psi\rangle\langle\psi|\hat{A}^\dagger = \hat{A}\hat{\rho}\hat{A}^\dagger.$$ (2.3)

Its Wigner-Weyl transform is then given by:

$$\begin{aligned} W[\hat{\rho}_A] &= W[\hat{A}\hat{\rho}\hat{A}^\dagger] = W[\hat{A}] \star W[\hat{\rho}] \star W[\hat{A}^\dagger] \\ &= W[\hat{A}] \star W[\hat{\rho}] \star W[\hat{A}]^* = A(q,k) \star W(q,k) \star A(q,k)^*, \end{aligned}$$ (2.4)

where the property in Eq. (1.15) was used and A^* denotes the complex conjugation of A. This is the transformation of the transform of a state ψ when an operator acts upon it. This formulation is useful for NC QM since the Moyal $\star$-product is easily generalized to fit the additional commutation relations introduced in the theory, without sacrificing any of its fundamental properties. The NC QM is characterized by the deformation of the HW algebra as introduced in Chapter 1, Eq. (1.19), with the variables of the two algebras being connected by the so-called SW map Eq. (1.22). This allows for the generalization of most results directly from QM to NCQM with some adjustments, as was discussed in Chapter 1. This will provide a unified framework for the treatment of the 'No-cloning' and 'No-deleting' theorems which will be evident in the discussion in Sections 2.2 and 2.3.

2.2   Fidelity

The fidelity of two states measures the similarity between to states. It is given by:

$$F = \left(\mathrm{Tr} \left(\sqrt{\rho^{1/2} \sigma \rho^{1/2}} \right) \right)^2,$$ (2.5)

where ρ and σ are two density matrices corresponding to different quantum states. If one of the states is a pure one, for example $\rho = |\Psi\rangle\langle\Psi|$, then it reduces to:

$$F = \langle \Psi | \hat{\sigma} | \Psi \rangle,$$ (2.6)

and also $F = |\langle \Psi | | \phi \rangle |^2$ if $\sigma = |\phi\rangle \langle\phi|$. In the Wigner formalism, the fidelity can be written as:

$$F = \int dx \int dp \, W^{in}(x,p) \, W^{out}(x,p).$$ (2.7)

Here, the superscripts *in* and *out* are used to distinguish between the two different states. However, in the context of quantum teleportation (to be addressed later) they label the *input* and *output* states of this procedure. For the above process of ideal teleportation:

$$F = \int dx \int dp \, W^{in}(x,p)^2.$$ (2.8)

If the input state is a pure one, then the fidelity of this method is unity. However, if the shared entangled state is not a maximally entangled one then $F \leq 1$.

Finally, it should be pointed out that in the following sections we shall make use of phase-space formalism without resorting to Dirac's "bra-ket" notation, so to be compatible with the Weyl-Wigner formalism.

2.3 No-cloning theorem in the NC phase-space

In order to introduce the no-cloning theorem in the NC phase-space one will first review the well-known results of the no-cloning theorem in the Wigner formalism of phase-space QM.

First one considers two unknown states with Wigner functions $W_\psi(\mathbf{z})$ and $W_\phi(\mathbf{z})$ and, additionally, a blank third state, $W^e(\mathbf{z})$. All states are assumed to be normalized. Any cloning procedure should be able to take any unknown initial state (as well as an empty one) and replicate this state completely. In terms of QM one should be able to find a unitary transformation such that (cf. Eq. (2.4)),

$$W_\psi^{out}(\mathbf{z}^A, \mathbf{z}^B) = U^*(\mathbf{z}^A, \mathbf{z}^B) \star \left[W_\psi(\mathbf{z}^A) W^e(\mathbf{z}^B) \right] \star U(\mathbf{z}^A, \mathbf{z}^B) = W_\psi(\mathbf{z}^A) W_\psi(\mathbf{z}^B),$$ (2.9)

for any state ψ. One should notice that, for unitary transformations, $U^*(\mathbf{z}^A, \mathbf{z}^B) \star U(\mathbf{z}^A, \mathbf{z}^B) = 1$. This also implies that for these transformations $U(\mathbf{z})^{-1} = U(\mathbf{z})^*$, where $U(\mathbf{z})^{-1}$ denotes the $\star$-product inverse. If one applies this procedure to the aforementioned states labeled by ψ and ϕ one can then compute overlap of these states using Eq. (2.2) by integrating:

$$P = \int d\mathbf{z}^A \int d\mathbf{z}^B \, W_\psi^{out}(\mathbf{z}^A, \mathbf{z}^B) \, W_\phi^{out}(\mathbf{z}^A, \mathbf{z}^B).$$ (2.10)

Indeed it can be carried out in two different ways, as illustrated in the following. First, by using the last equality on Eq. (2.9) one gets:

$$P = \int dz^A \int dz^B \, W_\psi(z^A) W_\psi(z^B) \, W_\phi(z^A) W_\phi(z^B)$$
$$= \int dz^A \, W_\psi(z^A) \, W_\phi(z^A) \int dz^B \, W_\psi(z^B) \, W_\phi(z^B) \tag{2.11}$$
$$= \left(\int dz \, W_\psi(z) \, W_\phi(z) \right)^2 .$$

However, if one uses the first equality of Eq. (2.9) and the associativity of the $\star$-product, the integral is written as:

$$P = \int dz^A \int dz^B \, U^*(z^A, z^B) \star \left[W_\psi(z^A) W^e(z^B) \right] \star U(z^A, z^B)$$
$$U^*(z^A, z^B) \star \left[W_\phi(z^A) W^e(z^B) \right] \star U(z^A, z^B)$$
$$= \int dz^A \int dz^B \, U^*(z^A, z^B) \star \left[W_\psi(z^A) W^e(z^B) \right] \star U(z^A, z^B) \star \tag{2.12}$$
$$\star U^{-1}(z^A, z^B) \star \left[W_\phi(z^A) W^e(z^B) \right] \star U(z^A, z^B)$$
$$= \int dz^A \int dz^B \, U^*(z^A, z^B) \star \left[W_\psi(z^A) W^e(z^B) \right] \star \left[W_\phi(z^A) W^e(z^B) \right] \star U(z^A, z^B),$$

where one has used the identity, Eq. (1.67) and the unitarity of $U(z^A, z^B)$. By following the $\star$-product associativity and the commutativity of the regular product, one obtains:

$$P = \int dz^A \int dz^B \, \left\{ U^*(z^A, z^B) \star \left[W_\psi(z^A) W^e(z^B) \right] \star \left[W_\phi(z^A) W^e(z^B) \right] \right\} U(z^A, z^B)$$
$$= \int dz^A \int dz^B \, U(z^A, z^B) \star U^{-1}(z^A, z^B) \star \left[W_\psi(z^A) W^e(z^B) \right] \star \left[W_\phi(z^A) W^e(z^B) \right] \tag{2.13}$$
$$= \int dz^A \int dz^B \, \left[W_\psi(z^A) W^e(z^B) \right] \star \left[W_\phi(z^A) W^e(z^B) \right] ,$$

which can be recast in the simplified form of:

$$P = \int dz^A \int dz^B \, \left[W_\psi(z^A) W^e(z^B) \right] \left[W_\phi(z^A) W^e(z^B) \right]$$
$$= \int dz^A \, W_\psi(z^A) \, W_\phi(z^A) \int dz^B \, W^e(z^B)^2 \tag{2.14}$$
$$= \int dz \, W_\psi(z) \, W_\phi(z).$$

Therefore, this cloning procedure implies that $P^2 = P$. Since P is a real number, i.e. the overlap probability, then the only possible solutions are that $P = 0$ or $P = 1$. This in turn implies that either the states are orthogonal or are the same state, which is absurd since it was assumed that the associated ψ and ϕ states were generic and unknown. This leads to the conclusion that such a cloning mechanism does not exist, proving the no-cloning theorem.

Most importantly, having the above derivation in mind, one should notice that the properties used there are only concerned with the $\star$-product defined for the theory. Thus, according to the results from Ref. [15] expressed by theorem Eq. (1.67), and the ensued invariance of Eq. (2.10) under the SW map, and by the associativity of the NC $\star_{NC}$-product, a proof as the above one can be extended to NC QM. In fact, this generalizes the no-cloning theorem to any theory where the $\star$-product obeys the two aforementioned properties: $\star_\Lambda$-product associativity and the theorem, Eq. (1.67)[1].

No-deleting theorem

A similar argument can also be used to prove the converse of the no-cloning theorem: the no-deleting theorem. In this case, one should look for a unitary transformation that yields:

$$W_\psi^{out}(\mathbf{z}^A,\mathbf{z}^B) = U^*(\mathbf{z}^A,\mathbf{z}^B) \star \left[W_\psi(\mathbf{z}^A)W_\psi(\mathbf{z}^B)\right] \star U(\mathbf{z}^A,\mathbf{z}^B) =$$
$$= W_\psi(\mathbf{z}^A)W^e(\mathbf{z}^B). \tag{2.15}$$

Considering two states, ψ and ϕ, and using the second equality we get:

$$\begin{aligned}
P &= \int d\mathbf{z}^A \int d\mathbf{z}^B \, W_\psi(\mathbf{z}^A)W^e(\mathbf{z}^B) \, W_\phi(\mathbf{z}^A)W^e(\mathbf{z}^B) \\
&= \int d\mathbf{z}^A \, W_\psi(\mathbf{z}^A) \, W_\phi(\mathbf{z}^A) \int d\mathbf{z}^B \, W^e(\mathbf{z}^B)^2 \\
&= \int d\mathbf{z} \, W_\psi(\mathbf{z}) \, W_\phi(\mathbf{z}).
\end{aligned} \tag{2.16}$$

---

[1] Some of the steps of this result have already been obtained in Ref. [41], although through a different procedure, which did not rely on the phase-space Wigner formalism for NC QM. The method employed here allows for the generalization to any deformation of the HW algebra.

However, if instead, the first equality is used, after some straightforward algebra, the result becomes:

$$
\begin{aligned}
P &= \int dz^A \int dz^B \, U^*(z^A, z^B) \star \left[W_\psi(z^A) W_\psi(z^B) \right] \star U(z^A, z^B) \\
&\quad U^*(z^A, z^B) \star \left[W_\phi(z^A) W_\phi(z^B) \right] \star U(z^A, z^B) \\
&= \int dz^A \int dz^B \, U^*(z^A, z^B) \star \left[W_\psi(z^A) W_\psi(z^B) \right] \star U(z^A, z^B) \star \\
&\quad \star U^{-1}(z^A, z^B) \star \left[W_\phi(z^A) W_\phi(z^B) \right] \star U(z^A, z^B) \\
&= \int dz^A \int dz^B \, U^*(z^A, z^B) \star \left[W_\psi(z^A) W_\phi(z^B) \right] \star \left[W_\phi(z^A) W_\phi(z^B) \right] \star U(z^A, z^B) \\
&= \int dz^A \int dz^B \, \left\{ U^*(z^A, z^B) \star \left[W_\psi(z^A) W_\psi(z^B) \right] \star \left[W_\phi(z^A) W_\phi(z^B) \right] \right\} U(z^A, z^B) \\
&= \int dz^A \int dz^B \, U(z^A, z^B) \star U^{-1}(z^A, z^B) \star \left[W_\psi(z^A) W_\psi(z^B) \right] \star \left[W_\phi(z^A) W_\phi(z^B) \right] \\
&= \int dz^A \int dz^B \, \left[W_\psi(z^A) W_\psi(z^B) \right] \star \left[W_\phi(z^A) W_\phi(z^B) \right] \\
&= \int dz^A \int dz^B \, \left[W_\psi(z^A) W_\psi(z^B) \right] \left[W_\phi(z^A) W_\phi(z^B) \right] \\
&= \int dz^A \, W_\psi(z^A) \, W_\phi(z^A) \int dz^B \, W_\psi(z^B) \, W_\phi(z^B) \\
&= \left(\int dz \, W_\psi(z) \, W_\phi(z) \right)^2 .
\end{aligned}
\tag{2.17}
$$

Thus, one concludes that $P = P^2$, which, by the same argument used on the QM no-cloning theorem, proves that there is no unitary transformation that acts on states as specified in Eq. (2.15). Thus, the no-deleting theorem follows. Again, this is valid for any deformation of the HW algebra that gives rise to a $\star_\Lambda$-product that is associative and obeys Eq. (1.67), namely in a particular framework of the NC QM.

2.4 Teleportation fidelity in the NC phase-space

The capability for teleporting quantum information is associated to the irreducible nonlocal content of QM, which is exemplified by the nonlocal features of an entangled quantum state [42]. Given the entanglement dynamics of two interacting subsystems of a composite quantum system [43], the teleportation of a single-mode quantum state can be engendered through a suitable variation of the original Einstein-Podolsky-Rosen (EPR) procedure [37, 43]. Once expressed in terms of continuous variable systems, the original formulation of the procedure brings about the nonlocal properties shared by two subsystems in the EPR state with perfect correlations in both position and momentum coordinates. These correlations can be expressed in the language of bipartite Gaussian states [44] for which the corresponding EPR phase-space Wigner function is written as [19]:

$$W_{EPR}(\mathbf{z}) = \frac{1}{\pi^2 \sqrt{\det[\Sigma_2]}} \exp\left(-\mathbf{z}^T \Sigma_2^{-1} \mathbf{z}\right), \tag{2.18}$$

where, in this case, $\mathbf{z} \equiv (x_1, p_1, x_2, p_2)$, which is associated to a set of orthogonal quadratures, for modes $\alpha_1 \equiv x_1 + i\, p_1$ and $\alpha_2 \equiv x_2 + i\, p_2$, and the covariance matrix, Σ_2, given by

$$\Sigma_2 = \frac{1}{2} \begin{pmatrix} \beta & \gamma \\ \gamma^T & \beta \end{pmatrix}, \tag{2.19}$$

where $\beta = \cosh(2r)\,\mathrm{Diag}[+1\ +1]$ represents the self-correlation of single subsystems and $\gamma = -\sinh(2r)\sigma_{(2)}^z = \sinh(2r)\,\mathrm{Diag}[-1\ +1]$ describes the correlation between the two subsystems, with both given in terms of the associated squeezing parameter, r.

Here, the real vector $\mathbf{z}$ defines the set of canonically conjugate variables, position and momentum, for the relevant pathways for a massive particle and quadrature amplitudes suitably associated to electromagnetic field modes. This yields:

$$W_{EPR}(\alpha_1, \alpha_2) = \frac{4}{\pi^2} \exp\left[-\cosh(2r)\left(|\alpha_1|^2 + |\alpha_2|^2\right) - 2\sinh(2r)\,\mathrm{Re}[\alpha_1 \alpha_2]\right]. \tag{2.20}$$

The entangled state, W_{EPR}, works as an auxiliary tool shared by *input* and *output* states, $\hat{\rho}_{in}$ and $\hat{\rho}_{out}$, for the construction of realistic teleportation protocols described by continuous variable quantum states [37, 42]. These are the basic engineering tools of teleportation protocols [45–47]. These protocols give support to the development of the convolutional formalism that will be considered in the construction of a NC version of a fidelity quantifier for teleported quantum states [48].

Turning the notation to the above mentioned single-mode continuous variable, α, the framework sets up the teleportation evolution described by [42]

$$\hat{\rho}_{out} = \int d\alpha \int d\alpha^* \, \hat{D}(\alpha)\, \hat{\rho}_{in}\, \hat{D}^\dagger(\alpha), \tag{2.21}$$

where $\hat{\rho}_{in}$ is the original teleported state and $\hat{D}(\alpha) = \exp\left(\alpha a^\dagger - \alpha^* a\right)$ is the displacement operator [38, 45, 46] such that $\hat{D}(\alpha)|\omega\rangle = |\omega + \alpha\rangle$. The quantitative measure of the reproducibility of the *input* state is prescribed by an *output* state, $\hat{\rho}_{out}$ (c.f. Eq. (2.21)), given by means of the entanglement fidelity [48]:

$$\mathcal{F}_E = \int d\alpha \int d\alpha^* \left| \mathrm{Tr}\left[\hat{D}(\alpha)\,\hat{\rho}_{in}\right] \right|^2, \tag{2.22}$$

which, in the context of phase-space variables, can be reconstructed in terms of Wigner functions.

The overall analysis, which includes finite (non-singular) degrees of correlation and incorporate inefficiencies in the measurement process [37], results into a continuous variable cloning protocol translated into a convolutional formalism [49, 50] statistically expressed by the correspondence between *input* and *output* states in terms of Wigner functions of quantum ensembles, W_{in} and W_{out}, related by

$$W_{out} = W_{in} \circ G_\sigma, \tag{2.23}$$

where $G_\sigma = \pi^{-2} \exp(-|\alpha|^2 \sigma)$ with $\sigma = \exp(-2r)$, and which, of course, can be extended to a multimodal framework through $\sigma \to \Sigma$, such that Σ is the covariance matrix.

According to the convolutional relation, Eq. (2.23), the entanglement fidelity between *input* and *output* n-modal Wigner functions can be written as

$$\mathcal{F}_\Sigma^W = (2\pi)^n \int d^n\mathbf{z}\, W_{out}(\mathbf{z}) W_{in}(\mathbf{z}) = (2\pi)^n \int d^n\mathbf{z} \int d^n\mathbf{z}'\, W_{in}(\mathbf{z}')\, G_\Sigma(\mathbf{z} - \mathbf{z}')\, W_{in}(\mathbf{z}), \tag{2.24}$$

with $\mathbf{z} \equiv (x_1, p_1, x_2, p_2, x_3, p_3, \ldots, x_n, p_n,)$, and

$$W_{out}(\mathbf{z}) = \int d^n\mathbf{z}'\, W_{in}(\mathbf{z}')\, G_\Sigma(\mathbf{z} - \mathbf{z}'), \tag{2.25}$$

where, in this case, the Gaussian cloner is given by

$$G_\Sigma(\mathbf{z}) = \frac{1}{\pi^n \sqrt{Det[\Sigma]}} \exp\left(-\mathbf{z}^T \Sigma^{-1} \mathbf{z}\right), \tag{2.26}$$

for an arbitrary Σ. For normalized *input* Wigner functions,

$$\int d^n\mathbf{z}\, W_{in}(\mathbf{z}) = 1, \tag{2.27}$$

and a n-partite Gaussian distribution, $W_{in}(\mathbf{z}) = G_\Sigma(\mathbf{z})$, one obtains the maximal value of the entanglement fidelity, which is given by $2^n / (3^n \sqrt{Det(\Sigma)})$. For pure states, with $\sqrt{Det(\Sigma)} = 1$, single-mode Gaussians exhibit optimal cloning fidelity, which leads to duplications of coherent Gaussian states with a fidelity of $2/3$ [38]. A unitary cloning transformation identified by the above Wigner function teleportation protocol provides two copies of a system with a continuous spectrum at the price of a non unity cloning fidelity where the cloner is identified by

$$\lim_{\Sigma \to \infty} G_\Sigma(\mathbf{z} - \mathbf{z}') = \delta^{(n)}(\mathbf{z} - \mathbf{z}'), \tag{2.28}$$

which, in this case, is reached by the implementation of an uncertainty principle (UP) violating protocol which returns

$$\lim_{\Sigma \to \infty} \mathcal{F}_\Sigma^W = (2\pi)^n \int d^n\mathbf{z}\, W_{in}^2(\mathbf{z}), \tag{2.29}$$

that is, the purity of the quantum state $W_{in}(\mathbf{z})$. For $W_{in}(\mathbf{z})$ identified by a Gaussian $G_\Gamma(\mathbf{z})$, one has

$$\lim_{\Sigma \to \infty} \mathcal{F}_\Sigma^\Gamma = (2\pi)^n \int d^n\mathbf{z}\, W_{in}^2(\mathbf{z}) = (2\pi)^n \frac{1}{(2\pi)^n \sqrt{Det[\Gamma]}} = \frac{1}{\sqrt{Det[\Gamma]}}, \tag{2.30}$$

the Gaussian purity.

In fact, the result from Eq. (2.24) for the hypothesis, Eq. (2.26), can be straightforwardly generalized to Gaussian states given by $G_\Gamma(\mathbf{z})$. In this case, $W_{in}(\mathbf{z}) = G_\Sigma(\mathbf{z})$, and with some involved mathematical manipulations, leads to [51, 52]:

$$\mathcal{F}_\Sigma^\Gamma = (2\pi)^n \int d^n\mathbf{z} \int d^n\mathbf{z}'\, G_\Gamma(\mathbf{z}')\, G_\Sigma(\mathbf{z} - \mathbf{z}')\, G_\Gamma(\mathbf{z}) = \frac{2^n}{\sqrt{Det[2\Gamma + \Sigma]}}. \tag{2.31}$$

In the context of the NC QM, for the case where $\mathbf{z}$ is a vector in the NC phase-space, which is related to the commutative one by the linear SW map given by $\tilde{\mathbf{z}} = \mathbf{S}\mathbf{z}$, a quite relevant result is that the covariance of a NC fidelity quantifier, $\mathcal{F}_\Sigma^W \to \mathcal{F}_\Sigma^{NC}$, under the SW map, i.e. an analogous quantifier in the commutative phase-space, where $\mathcal{F}_{\tilde{\Sigma}}^{\tilde{W}}$ will be obtained in the following.

According to the theorem Eq. (1.67) and following definition, Eq. (1.51) (*see also Eqs. (27)-(30) and Definition 3.8 in Ref. [15]*), a NC Wigner function written in terms of NC variables $\tilde{\mathbf{z}}$ is given by:

$$W^{NC}(\tilde{\mathbf{z}}) = \frac{1}{\sqrt{Det(\Omega)}} \tilde{W}(\mathbf{z}(\tilde{\mathbf{z}})) \equiv \frac{1}{\sqrt{Det(\Omega)}} \tilde{W}(\mathbf{S}^{-1}\tilde{\mathbf{z}}), \tag{2.32}$$

from which one can define a NC version of entanglement fidelity given by

$$\begin{aligned} \mathcal{F}_\Sigma^{NC} &= (2\pi)^n \int d^n\tilde{\mathbf{z}} \int d^n\tilde{\mathbf{z}}'\, W_{in}^{NC}(\tilde{\mathbf{z}}')\, G_\Sigma^{NC}(\tilde{\mathbf{z}} - \tilde{\mathbf{z}}')\, W_{in}^{NC}(\tilde{\mathbf{z}}) \\ &= \frac{(2\pi)^n}{Det(\Omega)} \int d^n\mathbf{z}' \left|\frac{\partial\tilde{\mathbf{z}}'}{\partial\mathbf{z}'}\right| \int d^n\mathbf{z} \left|\frac{\partial\tilde{\mathbf{z}}}{\partial\mathbf{z}}\right| \tilde{W}_{in}(\mathbf{z}'(\tilde{\mathbf{z}}'))\, G_\Sigma^{NC}\left(\tilde{\mathbf{z}}(\mathbf{z}) - \tilde{\mathbf{z}}'(\mathbf{z}')\right)\, \tilde{W}_{in}(\mathbf{z}(\tilde{\mathbf{z}})). \end{aligned} \tag{2.33}$$

Once that the Jacobian determinant $|\partial\tilde{\mathbf{z}}/\partial\mathbf{z}|$ is given by Eq. (1.25), for

$$\begin{aligned} G_\Sigma^{NC}(\tilde{\mathbf{z}} - \tilde{\mathbf{z}}') &= G_\Sigma^{NC}\left(\tilde{\mathbf{z}}(\mathbf{z}) - \tilde{\mathbf{z}}'(\mathbf{z}')\right) = G_\Sigma^{NC}\left(\mathbf{S}(\mathbf{z} - \mathbf{z}')\right) \\ &= \sqrt{\frac{Det(\tilde{\Sigma})}{Det(\Sigma)}} G_{\tilde{\Sigma}}(\mathbf{z} - \mathbf{z}') = \frac{1}{\sqrt{Det(\Omega)}} G_{\tilde{\Sigma}}(\mathbf{z} - \mathbf{z}'), \end{aligned} \tag{2.34}$$

where $\Sigma = \mathbf{S}\,\tilde{\boldsymbol{\Sigma}}\,\mathbf{S}^T$ and $\Omega = \mathbf{S}\,\mathbf{J}\,\mathbf{S}^T$, one has

$$
\begin{aligned}
\mathcal{F}_{\Sigma}^{NC} &= (2\pi)^n \int d^n\tilde{\mathbf{z}} \int d^n\tilde{\mathbf{z}}'\, W_{in}^{NC}(\tilde{\mathbf{z}}')\, G_{\Sigma}^{NC}(\tilde{\mathbf{z}} - \tilde{\mathbf{z}}')\, W_{in}^{NC}(\tilde{\mathbf{z}}) \\
&= \frac{(2\pi)^n}{\sqrt{Det(\Omega)}} \int d^n\mathbf{z} \int d^n\mathbf{z}'\, \tilde{W}_{in}(\mathbf{z}')\, G_{\tilde{\Sigma}}(\mathbf{z} - \mathbf{z}')\, \tilde{W}_{in}(\mathbf{z}) \\
&= \frac{1}{\sqrt{Det(\Omega)}}\, \mathcal{F}_{\tilde{\Sigma}}^{\tilde{W}},
\end{aligned}
\tag{2.35}
$$

which is independent from the SW map. The above expression for $\mathcal{F}_{\Sigma}^{NC}$ can be straightforwardly specialized to NC Gaussian states identified by $W_{in}^{NC}(\tilde{\mathbf{z}}) = G_{\Gamma}^{NC}(\tilde{\mathbf{z}})$, with Γ arbitrary. In this case, the expression for the NC version of the fidelity quantifier from Eq. (2.31) can be written as

$$
\begin{aligned}
\mathcal{F}_{\Sigma}^{\Gamma(NC)} &= (2\pi)^n \int d^n\tilde{\mathbf{z}} \int d^n\tilde{\mathbf{z}}'\, G_{\Gamma}^{NC}(\tilde{\mathbf{z}}')\, G_{\Sigma}^{NC}(\tilde{\mathbf{z}} - \tilde{\mathbf{z}}')\, G_{\Gamma}^{NC}(\tilde{\mathbf{z}}) \\
&= \frac{(2\pi)^n}{\sqrt{Det(\Omega)}} \int d^n\mathbf{z} \int d^n\mathbf{z}'\, G_{\tilde{\Gamma}}(\mathbf{z}')\, G_{\tilde{\Sigma}}(\tilde{\mathbf{z}} - \tilde{\mathbf{z}}')\, G_{\tilde{\Gamma}}(\mathbf{z}') \\
&= \frac{2^n}{\sqrt{Det(\Omega)}\sqrt{Det[2\tilde{\Gamma} + \tilde{\Sigma}]}} = \frac{2^n}{\sqrt{Det[2\Gamma + \Sigma]}} \\
&= \mathcal{F}_{\Sigma}^{\Gamma},
\end{aligned}
\tag{2.36}
$$

that is, for Gaussian states, the noncommutativity does not affect in any way the fidelity results for continuos variables teleportation protocols: a quantitative result that matches enhanced predictions of the NC phase-space version of the no-cloning theorem.

Teleportation fidelity in NC Harmonic Oscillator

Turning our attention to more complex quantum states, the above discussion can be extended to the quantum mechanical problem of the 2-dim NC harmonic oscillator (HO) [15, 53]. The quantum Hamiltonian on the NC $x_1 - x_2$ plane,

$$
\hat{H}_{HO}^{NC}(\hat{\tilde{\mathbf{z}}}) = \hat{\tilde{\mathbf{z}}} \cdot \hat{\tilde{\mathbf{z}}} \leftrightarrow \hat{H}_{HO}^{NC}(\hat{\mathbf{q}}, \hat{\mathbf{k}}) = \hat{\mathbf{q}}^2 + \hat{\mathbf{k}}^2 = \sum_{i=1,2} \hat{q}_i^2 + \hat{k}_i^2,
\tag{2.37}
$$

when re-written in terms of the commutative observables, $\hat{x}_i$ and $\hat{p}_i$, reads

$$
H_{HO}^{W}(\hat{\mathbf{x}}, \hat{\mathbf{p}}) = A^2\hat{\mathbf{x}}^2 + B^2\hat{\mathbf{p}}^2 + \gamma \sum_{i,j=1}^{2} \epsilon_{ij}\hat{p}_i\hat{x}_j,
\tag{2.38}
$$

where

$$A^2 \;\equiv\; \frac{\lambda^2}{2} + \frac{\eta^2}{8\mu^2\hbar^2}\,,$$

$$B^2 \;\equiv\; \frac{\mu^2}{2} + \frac{\theta^2}{8\lambda^2\hbar^2}\,,$$

$$\gamma \;\equiv\; \frac{1}{2\hbar}\,(\theta + \eta)\,, \tag{2.39}$$

and the constraint Eq. (1.23) guarantees that $4A^2B^2 = 1 + \gamma^2$.

The *stargen*value problem for the Hamiltonian from Eq. (2.38) reads

$$H_{HO}^W \star W_{n_1,n_2}(x,\mathbf{p}) = E_{n_1,n_2}\, W_{n_1,n_2}(x,\mathbf{p}), \tag{2.40}$$

which, from the analysis developed in Ref. [53], results into

$$W_{n_1,n_2}(x,\mathbf{p}) = \frac{(-1)^{n_1+n_2}}{\pi^2\hbar^2}\,\exp\left[-\left(\Omega_+ + \Omega_-\right)/\hbar\right] L_{n_1}^0\left(\Omega_+/\hbar\right) L_{n_2}^0\left(\Omega_-/\hbar\right), \tag{2.41}$$

where L_n^0 are the associated Laguerre polynomials, n_1 and n_2 are non-negative integers, and

$$\Omega_\pm = \frac{A}{B}x^2 + \frac{B}{A}\mathbf{p}^2 \mp 2\sum_{i,j=1}^{2}\left(\epsilon_{ij}p_ix_j\right), \tag{2.42}$$

such that the energy spectrum is given by

$$E_{n_1,n_2} = \hbar\left[2\alpha\beta(n_1 + n_2 + 1) + \gamma(n_1 - n_2)\right], \tag{2.43}$$

and one has

$$\int_{-\infty}^{+\infty}\!dx_1 \int_{-\infty}^{+\infty}\!dp_1 \int_{-\infty}^{+\infty}\!dx_2 \int_{-\infty}^{+\infty}\!dp_2\; \rho_{n_1,n_2}^W(x,\mathbf{p}) = 1. \tag{2.44}$$

The mapped variables $\Omega_\pm$ can be straightforwardly identified with $\Omega_\pm = |\alpha_1 \pm i\alpha_2|^2$ for

$$\alpha_j = q_j + ik_j, \quad j = 1, 2, \tag{2.45}$$

which shows that input NC Wigner functions described by Eq. (2.41) in the $\Omega_+ - \Omega_-$ plane are equivalent to the ordinary commutative Wigner functions with $\Omega_\pm = |\alpha_1 \pm i\alpha_2|^2$ given in terms of

$$\alpha_j = \sqrt{\frac{A}{B}}\,x_j + i\sqrt{\frac{B}{A}}\,p_j, \quad j = 1, 2, \tag{2.46}$$

which does not change the phase-space volume of integration and keeps valid the general result Eq. (2.30). By considering the Gaussian channel described by Eq. (2.20), NC and ordinary QM yield the same *output* states and, therefore, entanglement fidelity results are not affected.

2.5   Quantum teleportation

Quantum teleportation is a purely quantum phenomenon first discussed by Bennet *et al.*
[54] for discrete variable quantum systems. It allows for the transmission of a single particle
quantum state from one location to another via the exchange of two classical information bits
making use of the quantum entanglement. Due to these properties and from the fact that the
state of the teleported particle at the *input* location is destroyed, there is neither violation
of causality nor violation of the no-cloning theorem. The generalization of this result for
continuous variables was first put into place by Vaidman [55], in a setup similar to the one
of discrete variables. In the following analysis, the teleportation process is discussed in the
Wigner formalism, first for the usual QM and then in the NC framework.

2.5.1   Ideal teleportation – Standard QM case

In the first step in the preparation of this process, the two intervening parts must share an
entangled state of two particles, $W^{\mathrm{EPR}}\left(x^A, p^A, x^B, p^B\right)$, where $x^{A,B}$, $p^{A,B}$ correspond to the
position and momentum of particles belonging to Alice and Bob, respectively. Let the tele-
ported state be described by the Wigner function $W^{in}(x^C, p^C)$. Now one of the intervening
parts, Alice, gets particles A and C and Bob gets particle B. The complete state of the system
composed by the three particles is given by:

$$W(x^A, p^A, x^B, p^B, x^C, p^C) = W^{\mathrm{EPR}}\left(x^A, p^A, x^B, p^B\right) W^{in}(x^C, p^C). \tag{2.47}$$

Alice now uses a beam splitter on particles A and C so that their positions and momenta can
be written as:

$$x_\pm = \frac{1}{\sqrt{2}}(x^A \pm x^C), \qquad p_\pm = \frac{1}{\sqrt{2}}(p^A \pm p^C), \tag{2.48}$$

which, once introduced into the Wigner function argument yields:

$$W = W^{\mathrm{EPR}}\left(\frac{1}{\sqrt{2}}(x_+ + x_-), \frac{1}{\sqrt{2}}(p_+ + p_-), x^B, p^B\right) W^{in}\left(\frac{1}{\sqrt{2}}(x_+ - x_-), \frac{1}{\sqrt{2}}(p_+ - p_-)\right). \tag{2.49}$$

Since $[x_+, p_-] = 0$, Alice can now make a measurement of these two quantities. Then, the final state of particle B is given by:

$$
\begin{aligned}
W^{out}(x^B, p^B) &= W(x^B, p^B \mid x_+, p_-) \\
&= \int dx_- \int dp_+ \, W^{EPR}\left(\frac{1}{\sqrt{2}}(x_+ + x_-), \frac{1}{\sqrt{2}}(p_+ + p_-), x^B, p^B\right) \\
&\qquad\qquad\qquad \times W^{in}\left(\frac{1}{\sqrt{2}}(x_+ - x_-), \frac{1}{\sqrt{2}}(p_+ - p_-)\right) \quad (2.50)
\end{aligned}
$$

If the shared entangled state of particles A and B is a maximally entangled state, i.e. $W^{EPR} = \delta(x^A + x^B)\delta(p^A - p^B)$, then the final state of particle B is:

$$
W^{out}(x^B, p^B) = W^{in}\left(x^B + \sqrt{2}x_+, p^B - \sqrt{2}p_+\right). \tag{2.51}
$$

The next step in the teleportation process is the communication of Alice results to Bob. This corresponds to the value of x_+ and p_- and these are usually exchanged via classical communication. Then Bob proceeds with the adequacy of his state according to the information received, adjusting the position and momentum of particle B, i.e. $x^B + \sqrt{2}x_+ \to x^B$ and $p^B - \sqrt{2}p_+ \to p^B$, which leads to:

$$
W^{out}(x^B, p^B) = W^{in}(x^B, p^B). \tag{2.52}
$$

The above condition implies that the final state of particle B is the same as the initial state of particle C. Therefore, for any W^{in}, the particle B can be put into that state by the procedure described above. Since the particle B does not need to be in the same physical location as the other two particles, the state was teleported to its location.

This procedure can be straightforwardly generalized to more than one dimension as long as an EPR state can be prepared. This implies that $[x^i + x^j, p^i - p^j] = 0$ for $i \neq j$.

2.5.2 Ideal teleportation – NC QM case

For the NC scenario one should address to more than one dimension teleportation protocols. One then considers the same previous setup for three particles, now with position $(x_1^i, x_2^i) \equiv (x^i, y^i)$ and momentum $(p_1^i, p_2^i) \equiv (p_x^i, p_y^i)$, with $i, j = A, B, C$. The three particle Wigner function is then:

$$
W_{NC} = W_{NC}^{EPR}(x^A, y^A, p_x^A, p_y^A, x^B, y^B, p_x^B, p_y^B)W_{NC}^{in}(x^C, y^C, p_x^C, p_y^C). \tag{2.53}
$$

Using a beam splitter, in the same way as in the previous case, but in 2-dim, we introduce a new set of variables:

$$x_\pm = \frac{1}{\sqrt{2}}(x^A \pm x^C), \qquad y_\pm = \frac{1}{\sqrt{2}}(y^A \pm y^C), \qquad p_{x_\pm} = \frac{1}{\sqrt{2}}(p_x^A \pm p_x^C), \qquad p_{y_\pm} = \frac{1}{\sqrt{2}}(p_y^A \pm p_y^C),$$

(2.54)

which can be used to rewrite the Wigner function as in Eq. (2.49). The NC issues impose some constraints upon the quantities measured by Alice: now x_+ and y_+ cannot be measured simultaneously, since $[x_+, y_+] = 2i\theta \neq 0$. Nevertheless, it is possible to choose a set of four observables to measure simultaneously in a way that the final state of particle B matches the one of the initial particle, C. These observables are x_+, y_-, p_{x_-} and p_{y_+}, which constitute a complete set of commuting observables. The Wigner function describing the state of particle B after Alice's measurement is then given by:

$$W_{NC}^{out}(x^B, y^B, p_x^B, p_y^B) = \int dx_- \int dy_+ \int dp_{x_+} \int dp_{y_-} W_{NC}^{EPR} W_{NC}^{in}.$$

(2.55)

If, as in the ordinary QM teleportation, one assumes the shared state to be maximally entangled, i.e.

$$W_{NC}^{EPR} = \delta(x^A + x^B)\delta(y^A - y^B)\delta(p_x^A - p_x^B)\delta(p_y^A + p_y^B),$$

(2.56)

then the final state of particle B would be:

$$W_{NC}^{out} = W_{NC}^{in}(x^B + \sqrt{2}x_+, y^B - \sqrt{2}y_-, p_x^B - \sqrt{p_{x_-}}, p_y^B + \sqrt{2}p_{y_+}).$$

(2.57)

When Alice communicates her results on the measurements of x_+, y_-, p_{x_-} and p_{y_+}, Bob can make the necessary unitary transformations on particle B so to have:

$$W_{NC}^{out} = W_{NC}^{in}(x^B, y^B, p_x^B, p_y^B),$$

(2.58)

successfully teleporting the unknown original state of particle C into particle B.

Therefore, using the above theoretical setup, quantum teleportation of particle states with $F = 1$ is also possible in the framework of NC QM. Of course, the shared state used in this procedure is not experimentally viable and the results for the teleportation probabilities shall be also constrained by an EPR state as given by the preliminaries of Sec. IV, i.e. when squeezed Gaussian states are used as shared entangled states.

Chapter 3

Relativistic dispersion relation and putative metric structure in noncommutative phase-space

Lorentz symmetry is one of the foundational principles of modern physics and, in particular, of Special and General Relativity. However, in the attempt to reach a quantum theory of gravity, several proposed theories admit some violation of this symmetry, by means of a deformation of the relativistic dispersion relation (see Refs. [56–59] for reviews). This takes place, for instance, in solutions of string theory [60, 61] and in noncommutative field theories [13, 62, 63]. In a similar way, noncommutative (NC) quantum mechanics breaks this symmetry explicitly by introducing additional commutation relations, which provide an intrinsic momentum and length scale to the theory. As a consequence, this would lead to a deformation of the dispersion relation due to such scales. This may provide some useful insight into the deep consequences of NCQM.

As seen in Chapter 1, phase-space NCQM (PSNCQM) is characterized by the introduction of momentum-momentum and position-position non-vanishing commutation relations and may be regarded as a deformation of the standard Heisenberg-Weyl (HW) algebra of these observables, Eq. (1.19). Moreover, also as discussed in Chapter 1, quantum mechanics may alternatively be regarded in the Wigner-Weyl (WW) phase space formalism, which is equivalent to the standard operator-state formalism [28, 30, 31], and a noncommutative version of this formalism exists [15], also being equivalent to the deformation, , Eq. (1.19). Hence, one may view the NC deformation as modification of the associated Moyal $\star$-product and, consequently, of the underlying symplectic structure of the phase-space manifold, namely, of the symplectic form.

Thus, the introduction of the additional commutation relations in PSNCQM produces a deformation of this symplectic form, which captures all effects of NCQM through a modified Moyal product. This suggests that the effect of PSNCQM is to alter the geometric symplectic structure of the phase-space manifold and this will be used in the present work in order to compute the deformation in the relativistic dispersion relation. Our results suggest that there is a deformation of this relation that is suitable for phenomenological testing and, therefore, it might allow for constraining one of the noncommutative parameters of the theory.

The outline of this Chapter is as follows. In Section 3.1 the extended phase-space formalism is introduced and motivated. The content of this Section is inspired on the formalism of Refs. [64, 65]. The use of this formalism for the treatment of Lorentz transformations in phase-space (rather than in configuration space) is highlighted together with its use for obtaining relativistic Hamiltonians. In Section 3.2 the geometric implications of PSNCQM are introduced through the deformed symplectic form of the phase-space manifold. Together with the Darboux transformation, this is used to derive the deformed dispersion relation in the NC scenario. Additionally, it is shown that this deformation of the relativistic dispersion relation does not affect the speed of light for massless particles, in contrast with other deformations. Section 3.3 is devoted to an astrophysical test of the deformed relation via the use of gamma ray burst (GRB) data and an upper bound on a NC parameter is obtained. Finally, Section 3.4 concerns the introduction of a metric structure on the phase-space via a compatible triplet that admits a symplectic form, an almost complex structure (ACS) and a metric tensor.

3.1 Extended phase-space formalism

NCQM, likewise QM itself, is inherently non-relativistic, and thus it is not invariant under Lorentz transformations, but rather under Galilean transformations. However, it is logical to expect that its additional commutation relations imply some deformations of the relativistic dispersion relation [66, 67]. At first, one may think that since the HW algebra is deformed in a nonrelativistic context, it has little to do with relativistic physics. However, the introduction of the noncommutative parameters introduces in the theory a fundamental length and momentum scale. The existence of these scales is a fundamental aspect of the theory and can, in fact, be extended to the relativistic regime. In order to see this, we will not be directly concerned with PSNCQM itself, but rather with the deformation it introduces on the symplectic structure of the phase-space manifold. This discussion is based on the deformation caused in the standard symplectic structure of the WW formulation of quantum mechanics. Therefore, we shall study the geometric properties of the phase-space as a symplectic manifold and assess the consequences of deforming this structure. Furthermore, since quantum mechanical phase-space is still non-relativistic (as it treats time in a special way), an approach is needed to symplectic manifolds that extends the usual notion of phase-space to incorporate time as a coordinate on this manifold rather than an external parameter. For this, we shall introduce the notion of an extended phase-space in the following section. Subsequently, the extended phase-space, which can be shown to have Lorentz invariance under certain conditions, may be deformed and the consequence of this deformation is the breaking of Lorentz symmetry and hence a correction to the dispersion relation arises.

3.1.1   Construction

Consider the configuration space, a manifold, M, with coordinates $\{x^i\}$, $i = 1, 2, 3$, and T^*M, the cotangent bundle of M, with coordinates $z = \{x^i, p_i\}$. Thus, a symplectic manifold of dimension $2n$ can be built, (T^*M, ω), where ω is a symplectic form defined on T^*M. It is always possible to define the symplectic form (at least locally) as the exterior derivative of some one form, i.e.,

$$\omega = \mathrm{d}\chi. \tag{3.1}$$

Additionally, let us define a Hamiltonian, H, on T^*M, as $H : T^*M \to \mathbb{R}$. This leads to a Hamiltonian system, (T^*M, ω, H). If this Hamiltonian is an explicit function of time, i.e. $\partial_t H \neq 0$, then, it is useful to study the system in the manifold $T^*M \times \mathbb{R}$, with coordinates $\{p_i, x^i, t\}$, where the Poincaré-Cartan invariant form is defined as:

$$\chi' = \chi - H(p_i, x^i, t)\, \mathrm{d}t. \tag{3.2}$$

This structure on $T^*M \times \mathbb{R}$ does not lead to a symplectic structure, since all symplectic manifolds must be even dimensional. One may thus think of expanding this structure with an aditional coordinate, say p_0, and define an extended phase-space. The $2n + 1$-dimensional space $T^*M \times \mathbb{R}$ would be recovered under some on-shell condition for some function defined on the extended phase-space. This construction is indeed possible and a thorough explanation is provided in Refs. [64, 65]. An abridged version will be given in the following.

Consider a second manifold M_1 with coordinates $\{x_1^\mu\} = \{t, x^i\}$ and its tangent bundle T^*M_1 with coordinates $\{p_\mu^1, x_1^\mu\} = \{p_0, p_i, t, x^i\}$. This tangent bundle is known as the extended phase-space. A Hamiltonian, H_1, can now be defined on T^*M_1. However, since the objective is to describe a physical system on $T^*M \times \mathbb{R}$, some condition must be imposed on the extended Hamiltonian, H_1, and the appropriate coordinate p_0^1 must be defined. These requirements are met if one defines $p_0^1 := -\mathcal{H}$, where $\mathcal{H}$ is the value of the non-extended Hamiltonian, H. One should note the distinction between $H = H(p_i, t, x^i)$, which is a function on $T^*M \times \mathbb{R}$, and $\mathcal{H}$ which is the value this function assumes on some phase-space point.

If one considers s as a parameter on T^*M_1 (in the same way that t is a paramenter on T^*M for autonomous systems), then $\mathcal{H} = \mathcal{H}(s) = H(p_i(s), t(s), x^i(s)) \in \mathbb{R}$. Since it only depends on the parameter s, then it is a suitable coordinate for p_0^1. As for the condition on H_1, one demands that $H_1 = 0$ generates a $(2n+1)$-dimensional hypersurface on T^*M_1 that matches the Hamiltonian system in $T^*M \times \mathbb{R}$. Thus, one can define a map $\Psi: T^*M \times \mathbb{R} \to T^*M_1$ such that:

$$\{t, x^i, p_i\} \to \{t, x^i, \mathcal{H}, p_i\} := \{x_1^\mu, p_\mu^1\}. \tag{3.3}$$

The inverse map, $\Psi^{-1} : T^*M_1 \to T^*M \times \mathbb{R}$, takes us back to the original system by replacing all instances of the $\mathcal{H}$ coordinate with the Hamiltonian H. The symplectic form on T^*M_1 is, in canonical coordinates, given by:

$$\omega_e = dp_\mu^1 \wedge dx_1^\mu = dp_i \wedge dx^i - d\mathcal{H} \wedge dt. \tag{3.4}$$

If one applies Ψ^{-1} to this form, one gets:

$$\omega_{e|\mathcal{H}=H} = d\chi', \tag{3.5}$$

where χ' is the Poincaré-Cartan invariant defined in $T^*M \times \mathbb{R}$ in Eq. (3.2).

This construction of an extended phase-space that encompasses the tangent bundle of a four dimensional configuration space allows, not surprisingly, to study relativistic systems using the Hamiltonian formalism. This is particularly useful for us, since the noncommutative relations of NCQM can be encoded as a deformation of the standard symplectic form defined on the phase-space for a single particle.

3.1.2   Canonical transformations and Lorentz transformations

Consider the diffeomorphisms $\phi : T^*M_1 \to T^*M_1$ that preserve the symplectic form ω_e: these are defined as canonical transformations. The action of the pullback, ϕ^*, on the symplectic form is therefore $\phi^*\omega_e = \omega_e$. Since any symplectic form is, at least locally, exact (i.e., $\omega_e = d\chi_e$), one gets that the pullback of χ_e through ϕ must differ from χ_e itself by at most an exact 1-form, i.e. dF_1. Therefore, this function F_1 generates the coordinate transformations that preserve the symplectic structure of the manifold. It is possible to find a generating function, call it F_2, whose associated coordinate transformations are precisely the Lorentz transformations of the four-position and four-momentum vectors [64], thus allowing for the construction of Lorentz invariant Hamiltonians in the extended phase-space. Additionally, these transformations are not separable into a spatial coordinate transformation and a time scale change, thus having no analogous transformation in the regular phase-space. One can then take advantage of these transformations to find extensions for some non-relativistic

Hamiltonians, which become relativistic in the extended phase-space, and getting the relativistic Hamiltonian on T^*M using Ψ^{-1}.

3.1.3 Relativistic free particle

It is particularly useful to understand the discussed formalism, to consider the free particle non-relativistic Hamiltonian, given by:

$$H_{NL} = \frac{1}{2m}p^2, \tag{3.6}$$

which is defined in T^*M. One must now find an extension of H_{NL} such that it is Lorentz invariant in T^*M_1. This is straightforward and is given by:

$$H_1 = \frac{1}{2m}\left(p^2 - \frac{\mathcal{H}^2}{c^2}\right) + \frac{1}{2}mc^2, \tag{3.7}$$

where $\mathcal{H}$ is the deformation that extends the Hamiltonian to the extended space. This is properly defined in T^*M_1 and since it only depends on $(p^1)^2$ it is Lorentz invariant. The additional mc^2 term was added in order to have $H_L = mc^2$ when $p = 0$ in the regular phase-space. One should notice that the extended Hamiltonian H_1 is constructed as an extension of H_{NL}, in the sense that $H_{1|_{\mathcal{H}=0}} = H_{NL} + \text{constant}$. However, it is not a straightforward extension in the sense that, after the reduction process we recover H_{NL}: the resulting Hamiltonian should be Lorentz invariant. Indeed, if one now applies Ψ^{-1} to the system (T^*M_1, ω_e, H_1) it yields (T^*M, ω, H_L) with H_L being obtained by solving $H_1 = 0$ and replacing $\mathcal{H}$ by H_L, which leads to:

$$0 = \frac{1}{2m}\left(p^2 - \frac{H_L^2}{c^2}\right) + \frac{1}{2}mc^2 \tag{3.8}$$
$$\Leftrightarrow H_L^2 = p^2c^2 + m^2c^4,$$

which is the Hamiltonian for a relativistic free particle and yields the dispersion relation:

$$E^2 = p^2c^2 + m^2c^4 \tag{3.9}$$

3.1.4 Extended Phase-space and Noncommutativity

In order to find the deformation on the dispersion relation for a single relativistic particle due to the introduction of the additional commutation relations, one uses the extended phase-space formalism, as it is suitable to describe the phase-space of a relativistic particle. However, before the reduction process, one deforms the symplectic form ω_e in order to accommodate the additional commutation relations, and subsequently one has to find an appropriate coordinate change to bring the deformed form to its canonical form. For this purpose, one can make use of the Darboux's theorem, which guarantees that locally, any symplectic form is reducible to the standard one, by a change of coordinates.

3.2 Noncommutative Quantum Mechanics - symplectic structure

NCQM can be cast in the phase-space formalism of QM by deforming the symplectic form with the extra noncommutative parameters. This implies that the energy, i.e., Hamiltonian, will be different in this context. Before studying the consequences of this deformation, one needs to know how the symplectic form is affected by the introduction of the extra commutation relations. This is addressed in Subsection 1.1.3 and will be reviewed here in a brief manner. First one considers that the commutators are related with the symplectic form through (see Eqs. (1.37) and (1.43)):

$$[f, g] = i\hbar\omega\left(X_f, X_g\right),\tag{3.10}$$

where $f = f(\hat{z})$ and $g = g(\hat{z})$ are functions of position and momentum operators and $X_{f/g}$ are the vector fields associated with the corresponding function. This vector field is defined through the relation:

$$\iota_{X_f}\omega = -df,\tag{3.11}$$

where ι_X is the interior product of a vector field and a differential form, $\iota_X : \Omega^p \rightarrow \Omega^{p-1}$, defined by $(\iota_X\omega)(X_1, \ldots, X_n) = \omega(X, X_1, \ldots, X_n)$. In components:

$$\omega_{ij}X_f^i = -\partial_j f \Leftrightarrow X_f^i = -\omega^{ij}\partial_j f,\tag{3.12}$$

where $\omega^{ij} := \left(\omega^{-1}\right)_{ij}$ represents the components of the inverse of ω. When applied to the commutator this yields:

$$[f, g] = i\hbar\omega_{ij}X_f^i X_g^j = i\hbar\omega_{ij}\omega^{ik}\omega^{jl}\partial_k f\partial_l g = i\hbar\omega^{kl}\partial_k f\partial_l g.\tag{3.13}$$

In particular, this implies that:

$$\left[\hat{z}^i, \hat{z}^j\right] = i\hbar\omega^{ij}.\tag{3.14}$$

Therefore, the commutation relations determine uniquely the components of the inverse of the symplectic form. Given this result, one is able to write it in matrix form for the commutative scenario as:

$$\omega^{-1} = \begin{pmatrix} 0 & \mathbf{Id} \\ -\mathbf{Id} & 0 \end{pmatrix},\tag{3.15}$$

and for the noncommutative one has:

$$\omega_{NC}^{-1} = \begin{pmatrix} \frac{\Theta}{\hbar} & \mathbf{Id} \\ -\mathbf{Id} & \frac{N}{\hbar} \end{pmatrix}, \tag{3.16}$$

where $\mathbf{Id}$ is the $n \times n$ identity matrix and Θ and $\mathbf{N}$ are skew-symmetric matrices encoding noncommutativity with constant entries θ and η respectively (see e.g. Ref. [15]). The symplectic form is then given by:

$$\omega_{NC} = \left(1 - \frac{\eta\theta}{4\hbar^2}\right) \left[\left(4\hbar^2 - \eta\theta\right) dq^i \wedge dk_i - 4\eta\hbar \sum_{\substack{i,j=1 \\ i<j}}^{3} dq^i \wedge dq^j - 4\theta\hbar \sum_{\substack{i,j=1 \\ i<j}}^{3} dk_i \wedge dk_j \right]. \tag{3.17}$$

In this setup, one is able to find a global coordinate transformation that reduces the ω_{NC} to the standard form. Let the new coordinates be (x^i, p_i). The symplectic form is then written as:

$$\omega = -dx^i \wedge dp_i, \tag{3.18}$$

or in the matrix form as ω^{-1} in Eq. (3.15). As seen, in Ref. [15], a suitable change of coordinates for this task is given by Eq. (1.22). This coordinate transformation is noncanonical, since it does not preserve the form of ω. In the following section we will explore the consequences of deforming the spatial part of the symplectic form on the dispersion relation for a relativistic particle, using the extended phase-space formalism.

3.2.1 Deformed dispersion relation

We now consider a free non-relativistic particle with the same Hamiltonian given by Eq. (3.6) and follow the same procedure that defines the extended Hamiltonian, Eq. (3.7). Then, we consider that the symplectic form in these coordinates, (p_i, q^i), is the deformed NC one. Therefore, in order to obtain the dispersion relation for commutative coordinates, the coordinate change in Eq. (1.22) is used. This yields the extended Hamiltonian:

$$H_1 = \frac{1}{2m} \left(\left(p_i + \frac{\eta}{2\hbar}\epsilon_{ij}x^j\right)^2 - \frac{\mathcal{H}^2}{c^2} \right) + \frac{1}{2}mc^2. \tag{3.19}$$

This change of coordinates is non-canonical, as the form of the symplectic form is changed. This in turn implies that the Lorentz invariance of the final Hamiltonian is broken. Therefore, the introduction of the noncommutative parameters leads to a violation of Lorentz invariance. As a first consequence, this means that the ensuing results are not independent of the observer performing measurements on the energy and speed of particles. The consequences of this fact will be further explored in the next section. After the reduction process, the Hamiltonian for a free particle in the deformed phase-space is given by:

$$H_{NC}^2 = \left(p_i + \frac{\eta}{2\hbar}\epsilon_{ij}x^j\right)^2 c^2 + m^2 c^4, \tag{3.20}$$

which, interpreting H_{NC} as the measured energy of a free particle, leads to a modified relativistic dispersion relation:

$$E^2 = \left(p_i + \frac{\eta}{2\hbar}\epsilon_{ij}x^j\right)^2 c^2 + m^2 c^4. \tag{3.21}$$

In turn, this deformation yields a modification to the propagation of free massless particles, given by:

$$c' = \nabla_p E = \frac{1}{2E}\nabla_p E^2 = \frac{c^2}{E}\left(p_i + \frac{\eta}{2\hbar}\epsilon_{ij}x^j\right)\hat{x}^j, \tag{3.22}$$

which leads to the measurable speed of light as $\|c'\| = c$. Therefore, even with the deformation in the dispersion relation, the speed of light is maintained.

3.3   Testing the deformation of the dispersion relation

In order to assess the validity of the deformation introduced in the dispersion relation, and since Lorentz invariance is no longer a symmetry of the system, one should be able to test quantitatively our result using tests of Lorentz invariance. Given that the resulting modification is dependent on the distance from the photon to the observer, gamma ray burst (GRB) tests of Lorentz invariance are suitable to examine this deformation as the distance scale is sufficiently large for the effects to be detectable. Furthermore, as the observations of these phenomena are made from Earth, it is convenient to use spherical coordinates, (r, Θ, Φ), for the deformed dispersion relation. In these coordinates, the dispersion relation becomes:

$$E^2 = c^2 p^2 + \frac{c^2 \eta}{\hbar} r \left[\frac{\eta}{2\hbar} r f(\Theta, \Phi) + p_\Phi (\sin\Phi + \cos\Phi) + p_\Theta g(\Theta, \Phi)\right], \tag{3.23}$$

where $f(\Theta, \Phi) = (1/2)(2 - \cos\Phi\sin 2\Theta + \sin 2\Theta \sin\Phi + \sin^2\Theta \sin 2\Phi)$ and $g(\Theta, \Phi) = \cos\Theta\cos\Phi - \cos\Theta\sin\Phi + \sin\Phi - \sin\Theta$. One immediately notices that the energy of the particle is not independent from Θ and Φ, which clearly breaks isotropy. This is not surprising since the commutation relations also break this property. If one assumes that the observed photons carry only radial momentum their energy is then given by:

$$E^2 = c^2 p^2 + \frac{c^2 \eta^2}{2\hbar^2} r^2 f(\Theta, \Phi). \tag{3.24}$$

This is a reasonable assumption, as the distance between the source and the receiver is very large compared to the radius of the Earth. In order to test the obtained expression for the energy of massless particles, an additional assumption is made, namely that the angular factor is a minor contribution to the energy deformation. One could also argue that phenomenologically the distribution of GRBs in the sky is isotropic. With this assumption, the dispersion relation becomes:

$$E \simeq cp\sqrt{1 + \frac{\eta^2}{2\hbar^2}\frac{r^2}{p^2}} \simeq cp\left(1 + \frac{\eta^2}{4\hbar^2}\frac{r^2}{p^2}\right)$$
$$\Leftrightarrow \frac{\Delta E}{E} \simeq \frac{\eta^2}{4\hbar^2}\frac{r^2}{p^2}.$$

$$(3.25)$$

One is now able to use GRB data to estimate the magnitude of the noncommutative parameter η. In order to do so, one assumes that the noncommutative correction is at most given by the uncertainty value in the measurement of the GRB energy. Therefore one needs the GRB energy, its uncertainty and the distance from Earth. Additionally, one uses that $E = pc + O(\eta)$, and $O(\eta) \ll 1$, to get:

$$\frac{\Delta E}{E} = \frac{\eta^2 c^2}{4\hbar^2}\frac{r^2}{E^2} \Leftrightarrow \eta = \frac{2\hbar}{cr}\sqrt{E\Delta E}.$$

$$(3.26)$$

Hence, one needs to know the detected energy of the emitted photons, its uncertainty, as well as the distance of the GRB. Data with this level of detail is not available at the moment, as far as we are aware. Instead, one uses average values for the energy in a certain detected band as a substitute for the photon values of these quantities. In order to calculate the distance of the GRB, the redshifts of these objects are used and the light-travel distance is computed. Data of some selected GRBs [68–71] is presented in Table 3.1.

Distances are computed from the redshift of associated counterparts in other wavelengths using the standard Λ-CDM model, with parameters $H_0 = 70\,km\,s^{-1}Mpc^{-1}$, $\Omega_m = 0.27$, $\Omega_\Lambda = 0.73$ and $k = 0$, and is given by:

$$d(z) = d_H \int_0^z \frac{dz'}{(1+z')E(z')},$$

$$(3.27)$$

where $d_H = c/H_0$ is the Hubble distance and

$$E(z) = \sqrt{\Omega_r(1+z)^4 + \Omega_m(1+z)^3 + \Omega_k(1+z)^2 + \Omega_\Lambda},$$

$$(3.28)$$

with $H(z) = H_0 E(z)$ being the Hubble factor at redshift z. The energy is taken to be related to the fluence measured in the BATSE standard $50 - 300$ keV energy band. Since the fluence

TABLE 3.1: Gamma Ray Burst data [68–71] and upper bound on $\sqrt{\eta}$

	Fluence (F) (erg/cm^2)	ΔF ($\times 10^{-8}$ erg/cm^2)	z	r ($\times 10^{25}$m)	$\sqrt{\eta}$ (eV)
180703A	7.8096×10^{-6}	2.3435	0.6678	5.86083	4.252×10^{-12}
171010A	3.3315×10^{-4}	5.1486	0.3285	3.50175	1.712×10^{-11}
170214A	8.2807×10^{-5}	6.3294	2.53	10.5876	7.319×10^{-12}
160625B	2.7614×10^{-4}	9.7813	1.406	8.71085	1.216×10^{-11}
160623A	2.1145×10^{-6}	3.9158	0.367	3.82187	4.319×10^{-12}
160509A	8.7354×10^{-5}	8.0344	1.17	8.02423	9.044×10^{-12}
150514A	2.5347×10^{-6}	2.4576	0.807	6.58455	3.064×10^{-12}
141028A	1.7317×10^{-5}	5.1850	2.33	10.3613	4.760×10^{-12}
140808A	2.1951×10^{-6}	1.6704	3.29	11.2211	2.056×10^{-12}
140801A	8.7936×10^{-6}	2.1775	1.32	8.47863	3.576×10^{-12}
140623A	1.6891×10^{-6}	2.5486	1.92	9.7792	2.293×10^{-12}
140620A	3.4297×10^{-6}	3.3173	2.04	9.96892	2.895×10^{-12}
140606B	3.7519×10^{-6}	2.4159	0.384	3.95833	4.341×10^{-12}
140508A	3.0887×10^{-5}	6.2225	1.027	7.5222	6.757×10^{-12}

is given by $F = E/A$, where A is the cross sectional area of detection, Eq. (3.26) changes to:

$$\eta = \frac{2\hbar A}{cr} \sqrt{F\Delta F}. \tag{3.29}$$

With the data above [68–71], the estimate on η is given, up to an area order of magnitude. This estimations are shown in Table 3.1 for $A = 1m^2$.

These results allows us to conclude that, if we take the uncertainty in the GRB energy as due to the deformed dispersion relation, given in Eq. (3.21), the magnitude of the fundamental momentum scale introduced by NCQM, $\sqrt{\eta}$, would be of the order of $\sim 10^{-12}$ eV/c for $A \sim 1\,\mathrm{m}^2$. This is a very stringent upper bound, improving on previous results where low-energy tests of Lorentz invariance were used, namely $\sqrt{\eta} \lesssim 10^{-5}$eV/c [24, 25].

3.4 Riemannian metric on noncommutative phase-space

It is useful for the understanding of the deformed dispersion relation, Eq. (3.21), to introduce a Riemannian metric in the phase-space, or more precisely, in the extended phase-space. In order to do this, one first considers a smooth manifold equipped with a differential 2-form, ω. This is a symplectic manifold if ω is closed and nondegenerate. This 2-form ω is called a symplectic 2-form. This structure is the basis of the phase-space.

The usual symplectic form defined on phase-space is $\omega = -dx^i \wedge dp_i$, $i = 1, \ldots, n$, which in matrix form is given by Eq. (3.15). In addition to the symplectic form, one can introduce an almost complex structure (ACS) on M. This is a smooth (1,1)-rank tensor, J such that $J^2 = -\mathrm{Id}$. A manifold equipped with an ACS is called an almost complex manifold.

Given a symplectic manifold (M, ω), it is always possible to find an ACS in such a way that a metric is well defined on the manifold [72]. In order to do so, one must find a compatible ACS. This condition is satisfied if:

$$\omega\left(J\left(X_{z^i}\right), J\left(X_{z^j}\right)\right) = \omega\left(X_{z^i}, X_{z^j}\right),$$

(3.30)

where X_{z^i} is the vector field associated with the function z^1, defined by Eq. (3.11). This condition thus requires that the ACS preserves the action of the symplectic form on vector fields in TM. Let us now consider the structure (M, ω, J) as a symplectic manifold equipped with an ACS. Then, it is always possible to define a Riemannian metric on M, i.e., a symmetric rank (0,2)-tensor, given by:

$$g\left(X_{z^i}, X_{z^j}\right) := \omega\left(X_{z^i}, J(X_{z^j})\right).$$

(3.31)

These objects form a compatible triplet, in the sense that defining two of these structures allows for specifying the third one in a unique way.

3.4.1 Commutative phase-space

For the commutative phase-space, the symplectic form is given by:

$$\omega = -dx^i \wedge dp_i,$$

(3.32)

$i = 1, ..., n$. The ACS satisfying Eq. (3.30) acts on the coordinate vector fiedls as:

$$J(X_{x^i}) = X_{p_i}, \quad J(X_{p_i}) = -X_{x^i}$$

(3.33)

The induced compatible Riemannian metric is then $g = \mathrm{Id}_{2n}$.

3.4.2 Noncommutative phase-space

The deformed symplectic form of the NC phase-space is easily characterized by its inverse as seen in Eq (3.16). The ACS action on the vector fields X_{z^i} is given by:

$$J(X_{q^i}) = \sqrt{\frac{\theta}{\eta}} X_{k_i}$$

$$J(X_{k_i}) = -\sqrt{\frac{\eta}{\theta}} X_{x^i}$$

(3.34)

In this scenario, the metric of the compatible trio acts on the above vector fields as:

$$g\left(X_{q^i}, X_{q^j}\right) = \sqrt{\frac{\theta}{\eta}}\,\delta_{ij}$$

$$g\left(X_{k_i}, X_{k_j}\right) = \sqrt{\frac{\eta}{\theta}}\,\delta_{ij}$$

$$g\left(X_{q^i}, X_{k_j}\right) = -\sqrt{\frac{\eta\theta}{\hbar^2}}\,\epsilon_{ij}$$

$$g\left(X_{k_i}, X_{q^j}\right) = \sqrt{\frac{\eta\theta}{\hbar^2}}\,\epsilon_{ij}.$$

(3.35)

The metric components are related to its value on the vector fields $X_{\tilde{z}^i}$ through:

$$g_{ij} = \omega_{ik}\omega_{jl}\, g\left(X_{\tilde{z}^k}, X_{\tilde{z}^l}\right),\tag{3.36}$$

which, for $n = 2$, i.e., a 4-dimensional phase-space, yields in matrix form:

$$g = \begin{pmatrix}
\hbar^2\dfrac{\sqrt{\frac{\eta}{\theta}}\hbar^2 - \sqrt{\eta^3\theta}}{(\theta\eta - \hbar^2)^2} & 0 & 0 & \dfrac{\hbar\sqrt{\eta\theta}}{\hbar^2 - \theta\eta} \\[2.2em]
0 & \hbar^2\dfrac{\sqrt{\frac{\eta}{\theta}}\hbar^2 - \sqrt{\eta^3\theta}}{(\theta\eta - \hbar^2)^2} & -\dfrac{\hbar\sqrt{\eta\theta}}{\hbar^2 - \theta\eta} & 0 \\[2.2em]
0 & -\dfrac{\hbar\sqrt{\eta\theta}}{\hbar^2 - \theta\eta} & \hbar^2\dfrac{\sqrt{\frac{\theta}{\eta}}\hbar^2 - \sqrt{\eta\theta^3}}{(\theta\eta - \hbar^2)^2} & 0 \\[2.2em]
\dfrac{\hbar\sqrt{\eta\theta}}{\hbar^2 - \theta\eta} & 0 & 0 & \hbar^2\dfrac{\sqrt{\frac{\theta}{\eta}}\hbar^2 - \sqrt{\eta\theta^3}}{(\theta\eta - \hbar^2)^2}
\end{pmatrix}.\tag{3.37}$$

The volume form associated with this metric is:

$$\text{volume}_g = \sqrt{|g|}\,\mathrm{d}^4\tilde{z} = \frac{\hbar^4}{\hbar^2 - \eta\theta}\,\mathrm{d}^4\tilde{z}.\tag{3.38}$$

This coincides with the induced volume by the symplectic form:

$$\text{volume}_\omega = \omega \wedge \omega = \frac{\hbar^4}{\hbar^2 - \eta\theta}\,\mathrm{d}^4\tilde{z},\tag{3.39}$$

which is to be expected as the manifold is Kähler. One should notice that it is not possible to take the naive commutative limit, i.e. $\theta, \eta \to 0$, since the metric becomes ill-defined. A way around is to consider $\theta/\eta = \alpha$ and then take the above limit with α fixed to get:

$$g = \text{diag}\left(\sqrt{\alpha^{-1}}, \sqrt{\alpha^{-1}}, \sqrt{\alpha}, \sqrt{\alpha}\right),\tag{3.40}$$

which yields the commutative limit iff $\alpha = 1$, that is $\theta = \eta$.

3.4.3 Extended Noncommutative phase-space

Let us apply the former procedure to the extended symplectic form, Eq. (3.4). The action of the ACS on the vector fields of the spatial coordinates is the same as the one given in Eq. (3.34), while its action on the time and energy vector fields is the noncommutative one, i.e.,

$$J(X_t) = X_{\mathcal{H}}, \quad J(X_{\mathcal{H}}) = -X_t. \tag{3.41}$$

We are then able to define the extended noncommutative phase-space metric by its action on the coordinate vector fields as before. Its action on the spatial vectors is the same as in Eq. (3.35) and on the remaining vectors it is given by:

$$g(X_t, X_t) = g(X_{\mathcal{H}}, X_{\mathcal{H}}) = -1. \tag{3.42}$$

From this we get that the metric components are given by:

$$g = \begin{pmatrix} -1 & 0 & 0 & 0 & 0 & 0 \\ 0 & \hbar^2 \dfrac{\sqrt{\frac{\eta}{\theta}}\hbar^2 - \sqrt{\eta^3\theta}}{(\theta\eta - \hbar^2)^2} & 0 & 0 & 0 & \dfrac{\hbar\sqrt{\eta\theta}}{\hbar^2 - \theta\eta} \\ 0 & 0 & \hbar^2 \dfrac{\sqrt{\frac{\eta}{\theta}}\hbar^2 - \sqrt{\eta^3\theta}}{(\theta\eta - \hbar^2)^2} & 0 & -\dfrac{\hbar\sqrt{\eta\theta}}{\hbar^2 - \theta\eta} & 0 \\ 0 & 0 & 0 & -1 & 0 & 0 \\ 0 & 0 & -\dfrac{\hbar\sqrt{\eta\theta}}{\hbar^2 - \theta\eta} & 0 & \hbar^2 \dfrac{\sqrt{\frac{\theta}{\eta}}\hbar^2 - \sqrt{\eta\theta^3}}{(\theta\eta - \hbar^2)^2} & 0 \\ 0 & \dfrac{\hbar\sqrt{\eta\theta}}{\hbar^2 - \theta\eta} & 0 & 0 & 0 & \hbar^2 \dfrac{\sqrt{\frac{\theta}{\eta}}\hbar^2 - \sqrt{\eta\theta^3}}{(\theta\eta - \hbar^2)^2} \end{pmatrix}, \tag{3.43}$$

with volume form being the same as in Eq. (3.38), which again coincides with the volume form induced by ω_e. Likewise in the non-extended NC phase-space, the limit $\eta, \theta \to 0$ is not well defined unless $\theta = \eta$.

This metric structure might have implications in high energy and statistical physics phenomena and clearly deserves a more in depth research.

Chapter 4

Collapsing Shells and Black Holes: a quantum analysis

It is well-known that General Relativity (GR), as a description for gravity, is not compatible with the quantum formalism that governs the remaining fundamental interactions of Nature. Therefore, a quantum theory of gravity is needed. One can attempt to quantize gravity directly from GR, in a procedure similar to the one that classical field theories have undergone. This treatment is realized through the ADM formalism [73–81] for GR, through a canonical quantization procedure that leads to the Wheeler-deWitt (WdW) equation [82–84]. This method gives origin to a functional differential equation for the metric components which, in the general case, does not exhibit neither analytical nor stable numerical solutions. Furthermore, the interpretation of such a functional framework and its comparison with the classical theory are not trivial. However, for a special class of metrics, the functional equations become differential equations, bearing much more tractable mathematics, and a Klein-Gordon like equation arises for the metric components. This simplification, often referred to as the minisuperspace approximation [84–87], provides a simple framework for quantizing some interesting classes of metrics. This method has been used to study the quantum dynamics of several geometries, like ones with $SO(3)$ invariance and cosmological models [86–91]. The quantum cosmology of the Kantwosky-Sachs (KS) metric was studied in the context of noncommutative quantum mechanics in Refs. [26, 27, 92]. For the BH problem, a map from the Schwarzschild to a KS metric is considered. This allows for the introduction of conjugate momenta for the metric components in a spherically symmetric metric so to yield a Hamiltonian, that is a sum of constraints, and consequently, a well defined WdW equation for the problem [26, 27, 93, 94]. In this context, one of the purposes of the BH quantization followed along this manuscript (c. f. Sec. 4.1) is to examine the collapse of a shell into a BH.

In the setting to be studied, the gravitational collapse, namely, the collapse of a thin shell, which acts as a boundary dividing two spacetimes, is the most natural problem to be addressed. The standard method for solving this class of problems was introduced by Israel [95] in his formulation of the junction conditions needed for the resulting spacetime to be compatible with Einstein field equations. Although it was first formulated for timelike and spacelike shells, it was later extended for the case of null shells [96, 97]. The use of this formalism is two-fold. On one side, it is possible to define the energy-momentum tensor on the shell and deduce what are the possible spacetimes to be matched: the choice of one of them completely determines the features of the other. On the other side, one can specify the spacetimes to be connected and use the junction conditions to define relevant the quantities on the shell. Most of the literature on the quantization of collapsing shells is only concerned with the first approach (see, for instance, Refs. [98, 99] and references therein). In this work the second approach is chosen so to allow for a straightforward definition of the shell quantities in terms of the metric function of the embedding spacetimes. The use of a null shell, instead of a timelike one, is related to the fact that, for null ones, the surface energy-momentum tensor is fully determined by the geometric properties of spacetime, i.e., by the metric field. Thus, no extra degrees of freedom are necessary and the quantization procedure is carried out only with the metric.

Upon quantization, a set of commutation relations must be imposed on the metric components and their conjugate momenta. Usually these relations follow directly from the Poisson brackets of the corresponding quantities at classical level, and can be implemented at quantum level through the Heisenberg-Weyl algebra. Following the general purpose of the book, the commutation relations considered upon quantization follow the deformation given by Eq. (1.19), which have been previously discussed in the literature, e.g, in Refs. [6, 7, 9–12, 15–18, 20, 21, 92, 100–102]. The application to the deformed quantization of GR solutions via WdW equation has been considered for the KS metric [92] and BH [26, 27, 94]. It was shown that the phase-sapce NC deformation of the algebra works as a regulator for the BH singularity at the origin [26]. It is, thus, interesting to study the effect of the deformed algebra in the context of the gravitational collapsing shells.

This Chapter is structured as follows. In Sec. 4.1, the GR geometrical construction is performed as to explicitly express the metrics on both sides of the shell. A convenient map parametrization is introduced to simplify the ensuing developments. In Sec. 4.2, the action for the spacetime with a shell is computed considering a discontinuity term in the Ricci scalar and a boundary term for null-like surfaces. In addition, the Hamiltonian for the system is deduced. In Sec. 4.3, the Wheeler-deWitt quantization procedure is adopted and the Hamiltonian constraint equation is solved in commutative and phase-space NC scenarios.

4.1   The Black hole-mass shell

Let us consider a null shell of mass, m, falling into a Schwarzschild BH of mass, M. There are two metrics in this system, one corresponding to the interior of the shell and the other related to the exterior. These can be written as:

$$ds_{in}^2 = -\left(1 - \frac{2M}{r}\right)dt^2 + \left(1 - \frac{2M}{r}\right)^{-1}dr^2 + r^2\left(d\theta^2 + \sin^2\theta\, d\phi^2\right),\tag{4.1a}$$

$$ds_{out}^2 = -\left(1 - \frac{2(M+m)}{r}\right)dt^2 + \left(1 - \frac{2(M+m)}{r}\right)^{-1}dr^2 + r^2\left(d\theta^2 + \sin^2\theta\, d\phi^2\right).\tag{4.1b}$$

Depending on the position of the shell, there can be two event horizons: one at $r = 2M$ and the other at $r = 2(m+M)$. Since one intends to analyze the behavior of the shell as it collapses into the interior region, in this work we will consider the shell located at $r < 2M$. This choice has two motivations. On one hand, if the shell was located outside the event horizon, the minisuperspace simplifications to the Wheeler-de Witt equation would not hold, as the metric functions would be a function of r, making the treatment of the problem very different. On the other, considering the shell inside the horizon is quite natural and allows for the study of its behaviour near the singularity and to verify if the quantum version avoids the singularity problem. This choice makes the temporal and radial components of both metrics in Eq. (4.1) negative and therefore one is able to rewrite them as:

$$ds_{in}^2 = -\left(\frac{2M}{t} - 1\right)^{-1}dt^2 + \left(\frac{2M}{t} - 1\right)dr^2 + t^2\left(d\theta^2 + \sin^2\theta\, d\phi^2\right),\tag{4.2a}$$

$$ds_{out}^2 = -\left(\frac{2(M+m)}{t} - 1\right)^{-1}dt^2 + \left(\frac{2(M+m)}{t} - 1\right)dr^2 + t^2\left(d\theta^2 + \sin^2\theta\, d\phi^2\right),\tag{4.2b}$$

where all the metric coefficients are positive.

This allows for mapping of each metric into a KS-type metric which, in the Misner parameterization, is given by:

$$ds^2 = -N^2(t)dt^2 + e^{2\sqrt{3}\beta(t)}dr^2 + e^{-2\sqrt{3}(\beta(t)+\Omega(t))}\left(d\theta^2 + \sin^2\theta\, d\phi^2\right).\tag{4.3}$$

This is accomplished by using the map given in Ref. [27, 94],

$$N^2 = \left(\frac{2M}{t} - 1\right)^{-1},\quad e^{2\sqrt{3}\beta} = \left(\frac{2M}{t} - 1\right),\quad e^{-2\sqrt{3}(\beta+\Omega)} = t^2,\tag{4.4}$$

so that the two metrics can be written as:

$$ds_{in}^2 = -N_+^2 dt^2 + e^{2\sqrt{3}\beta_+}dr^2 + e^{-2\sqrt{3}(\beta_++\Omega_+)}\left(d\theta^2 + \sin^2\theta\, d\phi^2\right),\tag{4.5a}$$

$$ds^2_{out} = -N^2_- \, dt^2 + e^{2\sqrt{3}\beta_-} \, dr^2 + e^{-2\sqrt{3}(\beta_-+\Omega_-)} \left(d\theta^2 + \sin^2\theta \, d\phi^2\right), \tag{4.5b}$$

where $N_\pm$, $\beta_\pm$ and $\Omega_\pm$ are functions of the t coordinate only. Using this procedure one is able to carry out the quantization of the above stablished geometrical system maintaining all the symmetries of the problem. Nevertheless, before proceeding, we must be aware of some subtleties of this map. First, the relationship between Ω and t is not injective, being symmetric around the point $t = M$, the value for which $\Omega(t)$ has a minimum. This brings up no condition on the above parameterization, however, as the the map should return into a single set (β, Ω) for any given value of t, and since the function $\beta(t)$ is injective, the parameterization is indeed fiducial. Moreover, given the aforementioned symmetry for $\Omega(t)$, the distinction between each point symmetrically distanced from $t = M$ is made by the $\beta(t)$ function, having positive values for $t < M$ and negative values for $t > M$. Thus, two points, for example $t_1 = M + \varepsilon$ and $t_2 = M - \varepsilon$ (for $\varepsilon \leq M$), mapped using Eq. (4.4), become $t_1 \rightarrow (\beta_1, \Omega_0)$ and $t_2 \rightarrow (\beta_2, \Omega_0)$. Consequently, the behaviour of the wave functions near the singularity is distinguishable from the behaviour at the horizon by the value of β. As will be seen, the absence of a dependence on β implies the behaviour is the same in both positions.

A final remark concerning this map is related to the minimum of $\Omega(t)$. This point is reached at $\sim -\ln(M^2)$ which is negative for $M > 1$. However, because this minimum is finite, one can, without loss of generality, consider $M \leq 1$ and analyze the results for $\Omega > 0$. These properties of the map $t \rightarrow (\beta, \Omega)$ will be relevant in the analysis of the resulting wave function in Sec. 4.3.

4.1.1 Thin Null Shell Collapse

In order to pursue the treatment of thin shells collapsing we need to consider a manifold, $\mathcal{M}$, divided into two regions, $\mathcal{M}^+$ and $\mathcal{M}^-$, each with its own metric, $g_{\alpha\beta+}$ and $g_{\alpha\beta}^-$, respectively. Let us also set coordinates $x^\alpha_\pm$ accordingly. The separation surface between the two regions of the manifold is called Σ and for the matters of this work it is considered to be null, i.e., the normal vector to the surface is everywhere null (for the case of timelike and spacelike surfaces see Ref. [103]). In the following work, several quantities will be discussed as well as their behaviour at the separation surface Σ, thus we introduce the notation:

$$[A] := A|_{V^+} - A|_{V^-}, \tag{4.6}$$

for any quantity A. The main goal is to study the conditions that allow us to join smoothly the two regions of the manifold so that the whole manifold is a solution to EFE. For this, let us define the coordinates in Σ, y^a so that a basis for this surface is given by:

$$(e_\pm)^\alpha_a := \frac{\partial x^\alpha_\pm}{\partial y^a}. \tag{4.7}$$

We now choose more suitable coordinates (λ, θ^A) where λ is the parameter of the surface generators and θ^A labels the generator. This allows for a separation of the basis vectors into two spacelike vectors and a null one:

$$e^\alpha_A = \frac{\partial x^\alpha}{\partial \theta^A}, \qquad e^\alpha_\lambda = \frac{\partial x^\alpha}{\partial \lambda} := k^\alpha, \tag{4.8}$$

respectively. By construction, k^α is tangent to the generators of the surface, but since this surface is null, it is also normal to the surface and normal to e^α_A, that is:

$$k^\alpha k_\alpha = 0, \qquad k_\alpha e^\alpha_A = 0. \tag{4.9}$$

This leads to a particular form of the metric induced in Σ:

$$g_{\alpha\beta}\, e^\alpha_a\, e^\beta_b = g_{\alpha\beta}\, e^\alpha_A\, e^\beta_B := \sigma_{AB}, \tag{4.10}$$

so that the invariant interval is:

$$ds^2_\Sigma = \sigma_{AB}\, d\theta^A d\theta^B. \tag{4.11}$$

In order to guarantee a well-defined geometry on the separation surface we must demand that:

$$[\sigma_{AB}] = 0. \tag{4.12}$$

So far we have 3 basis vector for the 4 dimensional space, so a fourth one is needed to complete the basis. For this we must choose a vector that does not belong to Σ, i. e., is linearly independent from the three vectors above. Since a vector in $\mathcal{M}$ has four components, three constraints are needed to define the vector. Thus, we choose a vector for each region of the manifold, $\mathcal{M}^+$ and $\mathcal{M}^-$, a vector, N^α_+ and N^α_- respectively, that obeys the relations:

$$N^\alpha N_\alpha = 0, \qquad N^\alpha k_\alpha = -1, \qquad N_\alpha e^\alpha_A = 0. \tag{4.13}$$

This vector is not a linear combination of the other three since, if we consider their inner product we have:

$$N_\alpha \left(\alpha_1 k^\alpha + \alpha^A\, e^\alpha_A \right) = \alpha_1 N_\alpha k^\alpha = -\alpha_1 \neq 0, \tag{4.14}$$

and so we have proven the independence of this vector. It is then possible to write the inverse metric, in each side of Σ using the basis vectors as:

$$g^{\alpha\beta} = -k^{\alpha}N^{\beta} - N^{\alpha}k^{\beta} + \sigma^{AB}e_A^{\alpha}e_B^{\beta}. \tag{4.15}$$

The next step, to find the junction conditions for the physical quantities on the shell we must introduce timelike congruence that intersects Σ at any angle, such that each curve only does so this surface once. Since this congruence is not unique we can chose one that satisfies these requirements and set a parameter τ in a way that $\tau = 0$ coincides with the point of intersection with Σ. This parameter can be interpreted as the proper time of particles following those curves. With this construction, $\mathcal{M}^{-}$ corresponds to $\tau < 0$ and $\mathcal{M}^{+}$ to $\tau > 0$. The vector tangent to the congruence is denoted to be $u_{\pm}^{\alpha}$. Since we have built a basis for $\mathcal{M}$ we can write this vector in the basis $\{N^{\alpha}, k^{\alpha}, e_A^{\alpha}\}$ as:

$$u^{\alpha} = c_1 N^{\alpha} + c_2^{A} e_A^{\alpha} + c_3 k^{\alpha}, \tag{4.16}$$

where $c_1 = -u^{\alpha}k_{\alpha}$, $c_2^{A} = u_{\alpha}e_A^{\alpha}$ and $c_3 = -u^{\alpha}N_{\alpha}$. It is essencial to ensure that the chosen congruence is smooth (since it can define the time coordinate of the system), therefore we demand that the tangencial components of the tangent vector, u^{α}, must be continuous. This leads to the conditions:

$$[-u_{\alpha}k^{\alpha}] = 0, \qquad [u_{\alpha}e_A^{\alpha}] = 0. \tag{4.17}$$

The transverse component of this vector can be discontinuous due to the surface's stress-energy tensor, which will be addressed later. Also, we must note that imposing these constraints plus the geodesic equation allow us to determine the tangent vector in one of the regions of $\mathcal{M}$ by knowing its form in the other region. Regarding this parameter as a scalar field we can describe the surface Σ with the equation:

$$\tau(x^{\alpha}) = 0, \tag{4.18}$$

and thus the normal vector obeys $k_{\alpha} \propto \nabla_{\alpha}\tau$ and taking the component of u_{α} in the direction of k_{α} from Eq. (4.16) we can write:

$$k_{\alpha} = -\left(-k_{\mu}u^{\mu}\right)\frac{\partial \tau}{\partial x^{\alpha}}, \tag{4.19}$$

which is relevant since it ensures that this vector is continuous across Σ. This relation will be useful for the following steps.

Having constructed the null shell and its coordinate systems and metric, we can now evaluate what happens to physical quantities when crossing the shell. The first to consider is the

Riemann tensor, which is computed using:

$$R^{\rho}_{\sigma\mu\nu} = \partial_{\mu}\Gamma^{\rho}_{\nu\sigma} - \partial_{\nu}\Gamma^{\rho}_{\mu\sigma} + \Gamma^{\rho}_{\mu\lambda}\Gamma^{\lambda}_{\nu\sigma} - \Gamma^{\rho}_{\nu\lambda}\Gamma^{\lambda}_{\mu\sigma}. \tag{4.20}$$

Given the construction of the manifold, we can write the metric as a sum of two parts as:

$$g_{\alpha\beta} = g^{+}_{\alpha\beta}\Theta\left(\tau\right) + g^{-}_{\alpha\beta}\Theta\left(-\tau\right), \tag{4.21}$$

which leads to Christoffel symbols of the form:

$$\Gamma^{\alpha}_{\beta\sigma} = \Theta\left(\tau\right)\left(\Gamma^{+}\right)^{\alpha}_{\beta\sigma} + \Theta\left(-\tau\right)\left(\Gamma^{-}\right)^{\alpha}_{\beta\sigma}. \tag{4.22}$$

Using this expression in Eq. (4.20) yields:

$$R^{\rho}_{\sigma\mu\nu} = \Theta\left(\tau\right)\left(R^{+}\right)^{\rho}_{\sigma\mu\nu} + \Theta\left(-\tau\right)\left(R^{-}\right)^{\rho}_{\sigma\mu\nu} + \delta\left(\tau\right)\left(R_{\Sigma}\right)^{\rho}_{\sigma\mu\nu}, \tag{4.23}$$

where,

$$\left(R_{\Sigma}\right)^{\rho}_{\sigma\mu\nu} := \left[\Gamma^{\rho}_{\sigma\nu}\right]n_{\mu} - \left[\Gamma^{\rho}_{\sigma\mu}\right]n_{\nu} = \left(-k_{\alpha}u^{\alpha}\right)^{-1}\left(\left[\Gamma^{\rho}_{\sigma\nu}\right]k_{\mu} - \left[\Gamma^{\rho}_{\sigma\mu}\right]k_{\nu}\right), \tag{4.24}$$

which is the term that manifests the discontinuity of the Riemann tensor at Σ. In the second step we used Eq. (4.19) and the fact that $n_{\alpha} = \nabla_{\alpha}\tau$ is the normal vector to the surface, according to Eq. (4.18). Since we requested that the metric is continuous at this surface and also that the coordinates are the same on both sides, the tangencial derivatives of the metric must be continuous, that is:

$$\left[g_{\alpha\beta,\gamma}\right]k^{\gamma} = 0, \qquad \left[g_{\alpha\beta,\gamma}\right]e^{\gamma}_{A} = 0. \tag{4.25}$$

Therefore the only possible discontinuity is in the derivative along N^{α}. We can look a the previous equation as being a collection of vectors, one for each metric component. Each of this vectors must have null contraction with both k^{α} and e^{α}_{A} and a finite value for the contraction with N^{α}. Recalling Eq. (4.13), it is easy to see that these vectors must be multiples of k^{α} and so we conclude that there is a tensor $\gamma_{\alpha\beta}$ such that:

$$\left[g_{\alpha\beta,\gamma}\right] = -\gamma_{\alpha\beta}k_{\gamma}. \tag{4.26}$$

The minus sign only guarantees that the component of the derivative is positive if has the same orientation as N^{α}. Finally, we can rewrite the discontinuous part of the Riemann tensor using $\gamma_{\alpha\beta}$ as:

$$\left(R_{\Sigma}\right)^{\rho}_{\sigma\mu\nu} = \frac{1}{-2k_{\alpha}u^{\alpha}}\left(\left(\gamma^{\rho}_{\nu}k_{\sigma} - \gamma_{\sigma\nu}k^{\rho}\right)k_{\mu} - \left(\gamma^{\rho}_{\mu}k_{\sigma} - \gamma_{\sigma\mu}k^{\rho}\right)k_{\nu}\right). \tag{4.27}$$

With this, we can compute the Ricci tensor and the scalar curvature (see Ref. [104]), and then using EFE, we conclude that the discontinuous term of the Riemann tensor is equivalent to the existence of a energy-momentum on the surface Σ, given by:

$$(T_\Sigma)^{\alpha\beta} = (-k_\alpha u^\alpha)^{-1} \delta(\tau) S^{\alpha\beta}, \tag{4.28}$$

where we use:

$$S^{\alpha\beta} = \frac{1}{16\pi} \left(k^\alpha \gamma^\beta_\mu k^\mu + k^\beta \gamma^\alpha_\mu k^\mu - \gamma^\mu_\mu k^\alpha k^\beta - \gamma_{\mu\nu} k^\mu k^\nu g^{\alpha\beta} \right). \tag{4.29}$$

It is thus possible to conclude that if $\gamma_{\alpha\beta}$ vanishes then all physical quantities are continuous at the surface.

At this point we can summarize the conditions required for smoothly joining two geometries in order for the manifold to be a solution of EFE. These conditions are called the Israel junction conditions for null surfaces and can be written as:

$$[\sigma_{AB}] = 0, \qquad [g_{\alpha\beta,\gamma}] = -\gamma_{\alpha\beta} k_\gamma, \tag{4.30}$$

where the γ tensor is closely related to the existence of a energy-momentum tensor at the separation surface.

4.2   The Action

The first step in the quantization of a geometrical system is to write its action. Since the system consists of two different spacetimes, in contact at a boundary defined by the shell, two difficulties arise: first, there is a boundary in the system (an external boundary for the inner spacetime and an internal boundary for the outer spacetime), which requires the inclusion of a Gibbons-Hawking-York like term into the action; second, there is a discontinuity in the Ricci curvature scalar, which does affect the Einstein-Hilbert part of the action [104, 105]. Therefore, the resulting action is of the form:

$$S = S_{\mathrm{EH}} + S_{\mathrm{boundary}} = \frac{1}{16\pi} \int \sqrt{-g}\, \mathrm{d}^4 x \left[R_- \Theta(u - u_0) + R_+ \Theta(-u + u_0) + \delta R \right] + S_{\mathrm{boundary}}, \tag{4.31}$$

where natural units, $c = G = \hbar = 1$, have been considered and u is a coordinate that defines the position of the shell, a hypersurface Σ in the embedding manifold $\mathcal{M} = \mathcal{M}^+ \cup \mathcal{M}^- \cup \Sigma$. The term δR is the result of the discontinuity in quantities computed on both sides of the shell, and the boundary term, S_{boundary}, shall include two contributions, since the shell acts as a boundary for the inner and outer spacetimes, with quantities defined at each boundary depending on the metric of the spacetime.

It is useful to introduce a change of coordinates in the metrics given by Eq. (4.5) $u = r - \zeta t^*$, where $\zeta = \pm 1$ and $dt^* := N e^{-\sqrt{3}\beta} dt$. The two values of ζ correspond to either a collapsing shell ($\zeta = -1$) or an expanding one ($\zeta = 1$). The metrics then take the form:

$$ds^2 = e^{\sqrt{3}\beta} du \left(e^{\sqrt{3}\beta} du + 2\zeta N \, dt \right) + e^{-2\sqrt{3}\Lambda} \left(d\theta^2 + \sin^2\theta \, d\phi^2 \right), \tag{4.32}$$

where the $\pm$ symbols have been dropped and the combination $\Lambda = \beta + \Omega$ is introduced. Of course, the two metrics in Eq. (4.5) are expressed in the form of Eq. (4.32). The system depends now on the three functions N, β and Λ, on each side of the shell. Also, in these coordinates, the shell is located at a surface of constant u, i.e. $u = u_0$, which simplifies the forthcoming analysis.

In what concerns the nature of the shell in our problem, the Israel boundary conditions formalism for null hypersurfaces, fixes the geometries to be glued together and completely determines their behaviour [104]. Hence, even though the description of the whole system is essentially geometrical, the shell in the hypersurface of constant u gives to the action a contribution proportional to $\delta(u - u_0)$. In fact, the degrees of freedom of the shell are those of the embedding spacetime. This treatment fully determines the shell in the embedded geometry without the need of any extra degree of freedom.

4.2.1   The discontinuity term - δR

In order to further proceed, we must endow the space with a more robust geometric structure. For this, a basis of vectors for each side of the shell is constructed. A particularly useful choice is the set composed by the normal vector, the holonomic vectors on Σ for θ and ϕ, and a fourth vector to be introduced shortly. The normal vectors to a null hypersurface are given by:

$$n_\pm^\mu = \chi_\pm^{-1}(t, u) \, g_\pm^{\mu\nu} \partial_\nu \Phi(u), \tag{4.33}$$

where $\Phi(u) = 0$ defines Σ, and $\chi_\pm$ is a normalization factor that can be chosen arbitrarily. Thus, in this scenario $\Phi(u) = u - u_0$, and so:

$$n_\pm^\mu = -\frac{\zeta e^{-\sqrt{3}\beta_\pm}}{N_\pm \chi_\pm} \partial_t. \tag{4.34}$$

From here on, it is assumed that the time coordinate in both sides of the hypersurface is the same, so that the whole manifold can be parameterized by the same temporal parameter[1].

<hr>

With this choice, a natural set of coordinates on Σ is (t,θ,ϕ) since those coincide if the two sides are connected at a constant value $u = u_0$. Thus, the holonomic vectors are given by:

$$e^{\mu}_{(A)} := \frac{\partial x^{\mu}}{\partial y^A} = \delta^{\mu}_A \partial_A, \tag{4.35}$$

for $A = \theta, \phi$ and by

$$e^{\mu}_{(t)} := \frac{\partial x^{\mu}}{\partial t} = -\partial_t, \tag{4.36}$$

for t. With these vectors, we can define the induced metric on the hypersurface Σ as $g_{ab} = e^{\mu}_{(a)} e^{\nu}_{(b)} g_{\mu\nu}$, which yields:

$$ds^2_{|\Sigma_\pm} = e^{-2\sqrt{3}\Lambda_\pm} \left(d\theta^2 + \sin^2\theta \, d\phi^2 \right). \tag{4.37}$$

For this hypersurface to possess a well-defined geometry, the induced metric defined on it must be unique. This allows for imposing the first condition on the metric fields as $\Lambda_{|\Sigma-} = \Lambda_{|\Sigma+}$. However, since $\Lambda_\pm$ only depends on the coordinate t and not on u, the junction condition is valid for any t and therefore for the whole manifold. Moreover, the coordinate t covers the entire manifold, thus $\dot{\Lambda}_+|_{u=u_0} = \dot{\Lambda}_-|_{u=u_0}$.

In order to complete the basis of vector fields, an auxiliary one must be introduced (one at each side of Σ), because $n^{\mu} \propto e^{\mu}_{(t)}$. This additional vector field, M^{μ}, must not be orthogonal to n^{μ}, otherwise it would be a vector on Σ. Additionally, it shall be assumed null and orthogonal to the other basis vectors, i.e.

$$M^{\mu} M_{\mu} = 0, \qquad M \cdot n = \varepsilon^{-1} \neq 0, \qquad M \cdot e_{(A)} = 0, \tag{4.38}$$

for $A = \theta, \phi$ and ε arbitrary. The vector is thus determined (up to the constant factor) as:

$$M^{\mu}_\pm = \frac{\chi_\pm(u,t)}{\varepsilon_\pm} \partial_u - \frac{\zeta e^{\sqrt{3}\beta_\pm} \chi_\pm(u,t)}{2\varepsilon_\pm N_\pm} \partial_t. \tag{4.39}$$

From now on, a complete basis of vector fields for both sides of the hypersurface Σ is established, namely $\left(n^{\mu}, M^{\mu}, e^{\mu}_{(\theta)}, e^{\mu}_{(\phi)} \right)$. It allows for defining the quantities intrinsic to Σ, which do not depend on their four-dimensional counterparts (although they may be related in some way). These are the intrinsic quantities that will be used to solve the problem. For this purpose, the four-vectors are projected into the basis onto Σ. The normal vector can be written as $n^{\mu} = n^a e^{\mu}_{(a)}$, where its components are:

$$n^a_\pm = \frac{\zeta e^{-\sqrt{3}\beta_\pm}}{N_\pm \chi_\pm(u,t)} \partial_t. \tag{4.40}$$

The distinction between the four-vector and the three-vector is made through their indices:

greek indices represent four-dimensional quantities, lowercase latin indices represent three-dimensional quantities and uppercase latin indices represent two-dimensional quantities, particularly θ and ϕ. For the transverse vector, the one-form M_μ is projected onto Σ to give:

$$M_a^\pm = M_\mu^\pm e_{(a)}^\mu = \frac{\zeta N_\pm e^{\sqrt{3}\beta_\pm} \chi_\pm(u,t)}{\varepsilon_\pm} \partial_t. \tag{4.41}$$

The requirement that the quantities n^a and M_a coincide on both sides of the shell determines a unique structure for Σ. This condition imposes restrictions on the normalization factors, that is:

$$\left. \frac{e^{-\sqrt{3}\beta_-}}{N_- \chi_-} \right|_{u=u_0} = \left. \frac{e^{-\sqrt{3}\beta_+}}{N_+ \chi_+} \right|_{u=u_0}, \tag{4.42}$$

and,

$$\left. \frac{N_- e^{\sqrt{3}\beta_-} \chi_-}{\varepsilon_-} \right|_{u=u_0} = \left. \frac{N_+ e^{\sqrt{3}\beta_+} \chi_+}{\varepsilon_+} \right|_{u=u_0}. \tag{4.43}$$

These two equations together require that $\varepsilon_- = \varepsilon_+$. This is a sensible result since $\varepsilon = M \cdot n = M_\mu n^\mu = M_a n^a$. Using the condition Eq. (4.42) throughout the calculations, it is possible to verify that the resulting term for δR does not depend on the chosen normalization. Furthermore, one can notice that, since the metric variables only depend on t, the condition imposed at $u = u_0$ is, in fact, valid everywhere. This implies that the relationship between the lapse functions, N_+ and N_-, given by Eq. (4.42) is valid everywhere.

Having defined all the quantities necessary to study the structure of the hypersurface as a submanifold, one proceeds to compute the transverse curvature, defined by [104]:

$$\mathcal{K}_{ab} = M_\mu e_{(a);\nu}^\mu e_{(b)}^\nu, \tag{4.44}$$

where the semicolon denotes covariant derivatives. This tensor quantity is used instead of the more common extrinsic curvature (K_{ab}) as it relies on the changes of the metric along its normal vector which, in the case of null surfaces, is not a quite suitable choice. This occurs since the normal vector is also part of the hypersurface and does not carry information about the embedding space, resulting in an extrinsic curvature that is continuous across null surfaces (and across Σ in particular). Hence, the transverse curvature, which relies on the change of the metric along the introduced transverse vector, is used instead [104]. This can be computed for both sides of the hypersurface, for which the non-vanishing components are:

$$\begin{aligned}
\mathcal{K}_{tt}^\pm &= -\frac{\zeta e^{\sqrt{3}\beta_\pm} \chi_\pm}{\varepsilon} \left(\dot{N}_\pm + \sqrt{3} N_\pm \dot{\beta}_\pm \right), \\
\mathcal{K}_{\theta\theta}^\pm &= \frac{\sqrt{3}\zeta e^{\sqrt{3}(\beta_\pm - 2\Lambda_\pm)} \dot{\Lambda}_\pm \chi_\pm}{2\varepsilon N_\pm}, \\
\mathcal{K}_{\phi\phi}^\pm &= \mathcal{K}_{\theta\theta}^\pm \sin^2(\theta).
\end{aligned} \tag{4.45}$$

The transverse curvature is a tensor defined only on Σ and it can be used to extract the energy-momentum tensor on this surface. For this, a new tensor, γ_{ab}, is defined in the following way:

$$\gamma_{ab} := 2\left[\mathcal{K}_{ab}\right],\tag{4.46}$$

also only on Σ. Additionally, in order to compute the discontinuity in the energy-momentum tensor, it is necessary to find an equivalent to an inverse induced metric. This cannot be done in the usual way, since the induced metric is degenerate. However, it is possible to construct a "pseudo-inverse" tensor, g_*^{ab}, which obeys [104]:

$$g^{\mu\nu} = g_*^{ab}\, e_{(a)}^{\mu}\, e_{(b)}^{\nu} + \varepsilon\, n^a\left(e_{(a)}^{\mu} N^{\nu} + e_{(a)}^{\nu} N^{\mu}\right).\tag{4.47}$$

For the present problem, this three dimensional tensor can be written as:

$$g_*^{ab} = \begin{pmatrix} 0 & 0 & 0 \\ 0 & e^{2\sqrt{3}\Lambda} & 0 \\ 0 & 0 & e^{2\sqrt{3}\Lambda}\sin^{-2}(\theta) \end{pmatrix},\tag{4.48}$$

where each non-vanishing entry is the inverse of the corresponding entry in the induced metric. With this object, the surface energy-momentum tensor is computed to be:

$$T_{|\Sigma}^{ab} = \frac{1}{16\pi}\left[-\left(\gamma_{cd}\, g_*^{cd}\right) n^a n^b - \left(\gamma_{cd}\, n^c n^d\right) g_*^{ab} + \left(g_*^{ac}\, n^b n^d + g_*^{bc}\, n^a n^d\right)\gamma_{cd}\right],\tag{4.49}$$

for which its non-vanishing components are:

$$T_{|\Sigma}^{tt} = \frac{\sqrt{3}\zeta}{8\pi\varepsilon}\left[\frac{e^{-\sqrt{3}\beta_-}\dot{\Lambda}_-}{N_-^3\chi_-} - \frac{e^{-\sqrt{3}\beta_+}\dot{\Lambda}_+}{N_+^3\chi_+}\right],$$

$$T_{|\Sigma}^{\theta\theta} = -\frac{\zeta e^{-\sqrt{3}(\beta_- - 2\Lambda_-)}\left(\dot{N}_- + \sqrt{3}N_-\dot{\beta}_-\right)}{8\pi\varepsilon N_-^2\chi_-} + \frac{\zeta e^{-\sqrt{3}(\beta_+ - 2\Lambda_+)}\left(\dot{N}_+ + \sqrt{3}N_+\dot{\beta}_+\right)}{8\pi\varepsilon N_+^2\chi_+},\tag{4.50}$$

$$T_{|\Sigma}^{\phi\phi} = \frac{T^{\theta\theta}}{\sin^2(\theta)}.$$

Since this tensor is defined on Σ, we can compute its trace using the induced metric on this hypersurface, which yields:

$$T_{|\Sigma} = -\frac{\zeta e^{-\sqrt{3}\beta_-}\left(\dot{N}_- + \sqrt{3}N_-\dot{\beta}_-\right)}{4\pi\varepsilon N_-^2\chi_-} + \frac{\zeta e^{-\sqrt{3}\beta_+}\left(\dot{N}_+ + \sqrt{3}N_+\dot{\beta}_+\right)}{4\pi\varepsilon N_+^2\chi_+}.\tag{4.51}$$

The construction so far, as well as the requirements on the quantities on both sides, have been based on the assumption that the geometrical setup is a solution of the Einstein field equations (EFE). In particular, the definition of the surface stress-energy tensor follows from the application of these equations as it can be seen in Refs. [104, 105]. Therefore, the surface scalar curvature can be obtained directly from the trace of the stress-energy tensor by taking the trace of the EFE. This leads to:

$$R_{|\Sigma} = -8\pi T_{|\Sigma},$$

(4.52)

which, for the particular geometry at hand, yields:

$$R_{|\Sigma} = \frac{2\zeta e^{-\sqrt{3}\beta_-}\left(\dot{N}_- + \sqrt{3}N_-\dot{\beta}_-\right)}{\varepsilon N_-^2 \chi_-} - \frac{2\zeta e^{-\sqrt{3}\beta_+}\left(\dot{N}_+ + \sqrt{3}N_+\dot{\beta}_+\right)}{\varepsilon N_+^2 \chi_+},$$

(4.53)

where Eq. (4.42) has been used in order to set the resulting curvature scalar in a more symmetric fashion. The discontinuity in the scalar curvature, δR, is then obtained as [104]:

$$\delta R = \varepsilon R_{|\Sigma}\chi(u,r)\delta(u-u_0),$$

(4.54)

resulting in the expression:

$$\delta R = \left(\frac{2\zeta e^{-\sqrt{3}\beta_-}\left(\dot{N}_- + \sqrt{3}N_-\dot{\beta}_-\right)}{N_-^2} - \frac{2\zeta e^{-\sqrt{3}\beta_+}\left(\dot{N}_+ + \sqrt{3}N_+\dot{\beta}_+\right)}{N_+^2}\right)\delta(u-u_0),$$

(4.55)

which, in fact, is independent of the choice of χ and ε. We are now able to introduce this expression into the action and, after integrating over θ and ϕ, the resulting action for Σ coming from the Einstein-Hilbert (EH) action yields:

$$S_{\text{EH}}^{\Sigma} = \frac{1}{2}\int_{\Sigma}\left[\frac{2\zeta e^{-2\sqrt{3}\Lambda_-}\left(\dot{N}_- + \sqrt{3}N_-\dot{\beta}_-\right)}{N_-} - \frac{2\zeta e^{-2\sqrt{3}\Lambda_+}\left(\dot{N}_+ + \sqrt{3}N_+\dot{\beta}_+\right)}{N_+}\right]dt.$$

(4.56)

4.2.2 The Boundary term

Additionally to the surface term given by Eq. (4.56) as discussed in the previous section, an action term related to the boundary of the manifolds must be included. For spacelike or timelike boundaries this is expressed in terms of the Hawking-Gibbons-York term, which is given by:

$$S_{\text{GHY}} = \frac{\epsilon}{8\pi}\int_{\partial V}d^3y\,\sqrt{h}K,$$

(4.57)

where K is the trace of the extrinsic curvature, $\epsilon = +1$ (-1) if ∂V is timelike (spacelike) and the y integration is over the coordinates at the boundary of the manifold. In the present problem, however, the boundary of the manifold is null-like, and thus this term is not well-defined, since, for example, the induced metric is singular and hence its determinant vanishes. In a recent work, an analogue term for null surfaces was proposed [106] and in another it is extended for any boundary comprising any number of timelike, spacelike and null boundary segments [107]. In the present scenario, the boundary is null, and so one may consider the approach of Ref. [106]:

$$S_{\text{boundary}} = \frac{1}{8\pi} \int_{\partial V} dy_A^3 \sqrt{-g} \chi_\pm (\Theta + \kappa),$$

(4.58)

where y_A are the coordinates on Σ, κ is the parameter that measures the "non-affinity" of the parameter on the null generators, $\lambda = t$, and Θ is defined as:

$$\Theta = q^{\mu\nu}\Theta_{\mu\nu}, \quad \Theta_{\mu\nu} = q_\mu^\sigma q_\nu^\rho n_{\rho;\sigma},$$

(4.59)

where $q_{\mu\nu}$ is the induced two dimensional metric on Σ, extended to four dimensions. This metric can be computed as:

$$q_{AB} = g^{ab} e^a_{(A)} e^b_{(B)} = \begin{pmatrix} e^{-2\sqrt{3}\Lambda_\pm(t)} & 0 \\ 0 & e^{-2\sqrt{3}\Lambda_\pm(t)} \sin^2(\theta) \end{pmatrix}, \quad A, B = \theta, \phi.$$

(4.60)

It is simple to obtain the inverse, q^{AB}, as a regular inverse matrix. Then, the extension to four dimensions is obtained by using the basis vectors:

$$q^{\mu\nu} = q^{AB} e^\mu_{(A)} e^\nu_{(B)} = \begin{pmatrix} 0 & 0 & 0 & 0 \\ 0 & 0 & 0 & 0 \\ 0 & 0 & e^{2\sqrt{3}\Lambda_\pm(t)} & 0 \\ 0 & 0 & 0 & e^{2\sqrt{3}\Lambda_\pm(t)} \sin^{-2}(\theta) \end{pmatrix}.$$

(4.61)

These are uniquely defined on Σ since $[\Lambda] = 0$.

The computation of $\Theta_{\mu\nu}$ is rather straightforward. However, one must compute two tensors, one for the border of the inner space and the other for the outer space. The non-vanishing terms of this tensor are then:

$$\Theta_{\theta\theta}^\pm = -\frac{\sqrt{3}\zeta e^{-\sqrt{3}(\beta_\pm + 2\Lambda_\pm)} \dot{\Lambda}_\pm}{N_\pm \chi_\pm},$$

$$\Theta_{\phi\phi}^\pm = \Theta_{\theta\theta}^\pm \sin^2(\theta).$$

(4.62)

It follows that its trace is given by:

$$\Theta_\pm = -\frac{2\sqrt{3}\zeta e^{-\sqrt{3}\beta_\pm} \dot{\Lambda}_\pm}{N_\pm \chi_\pm}.$$

(4.63)

Considering this result and the boundary action given in Eq. (4.58), it is straightfrorward to see that this term does not depend on the specific normalization chosen for the normal vectors to the hypersurface.

Regarding κ, it can be computed by evaluating if the parameter on the null generators is affine. This yields:

$$n^{\mu}n^{\nu}_{;\mu} = \kappa n^{\mu} \iff \left(\frac{\zeta e^{-\sqrt{3}\beta_{\pm}}\dot{\chi}_{\pm}}{N_{\pm}\chi_{\pm}^3},0,0,0\right) = \kappa\left(-\frac{1}{\chi_{+}},0,0,0\right), \tag{4.64}$$

thus κ is found to be:

$$\kappa = -\frac{\zeta e^{-\sqrt{3}\beta_{\pm}}\dot{\chi}_{\pm}}{N_{\pm}\chi_{\pm}^2}. \tag{4.65}$$

The first observation regarding this term is that, when substituted into Eq. (4.58), it seems to depend on the normalization chosen for the null vectors. However, this is not true once the boundary conditions are imposed.

4.2.3 Full boundary action

When all the factors that contribute to the boundary part of the action are taken into account, we can write:

$$S_{\text{boundary}} = \frac{1}{2}\int dt\left[\sqrt{-g_{+}}\chi_{+}\left(\varepsilon\delta R_{+} + 2\Theta_{+} + 2\kappa_{+}\right) - \sqrt{-g_{-}}\chi_{-}\left(\varepsilon\delta R_{-} + 2\Theta_{-} + 2\kappa_{-}\right)\right]. \tag{4.66}$$

We now apply the boundary condition, Eq. (4.42), to the above action. It turns out that, not only the dependence on the derivative of $\chi_{\pm}$ cancels out, but also that the whole boundary action vanishes. This implies that, at least in the particular setup used in this work, the boundary terms that arise from the discontinuity in the derivatives of the metric fields are cancelled out by the terms which must be added to the Einstein-Hilbert action in the case of null boundaries. As a result, only the bulk terms of the action are relevant for the problem.

We can wonder why this precise cancelation happens since it appears to imply that the two spacetime pieces, each with its own metric, do not influence on each other and that the analysis could be carried for each piece separately. First, it should be clear that the cancelation happens because a null boundary was inserted in each of the spacetimes and hence those boundaries were matched while maintaining the full spacetime as a solution of EFE. As will be seen in the upcoming treatment, the existence of this matching will lead to an Hamiltonian which is different from the one that would be obtained from a KS manifold alone. Second, it is natural that this separation occurs, solely based on the behavior on null geodesics. These geodesics are determined completely by their null nature, unlike timelike ones, which depend on the conserved quantities of the underlying spacetime. Therefore, for an observer inside (outside) the shell, the trajectory it follows is completely determined by the requirement that it must be null on that spacetime and does not depend on anything else. This implies that trajectories are determined apart from the boundary conditions, hence the separation. Nevertheless, this does not mean that it is possible to connect whatever geometry by mathcing null surfaces with each other, since consistent matching conditions cannot be found in general.

4.2.4   Full action and Equations of motion

Now that the action for the boundary terms has been determined and proven to vanish in our scenario, it is only necessary to determine the bulk terms of the action. Since the spacetimes have the same structure, i.e., the metric components are written in the same way, the form of the Lagrangian for the inner and outter spaces are the same. Taking a metric of the form Eq. (4.32), the bulk action is straightforward to compute:

$$S_{\text{bulk}} = \frac{1}{2} \int dt\, du \left(e^{\sqrt{3}\beta} N + \frac{e^{-\sqrt{3}(\beta+2\Omega)}}{N} (3\dot{\beta}^2 - 3\dot{\Omega}^2) \right). \tag{4.67}$$

This leads to the Hamiltonian obtained in Refs. [92] (a slight rescaling of the Ω variable leads to a different coefficient on the potential term). Since all the terms have been computed, they can be inserted into Eq. (4.31) resulting in the expression:

$$\begin{aligned}
S = \frac{1}{2} \int dt\, du & \left[\left(e^{\sqrt{3}\beta_-} N_- + \frac{e^{-\sqrt{3}(\beta_-+2\Omega_-)}}{N_-} (3\dot{\beta}_-^2 - 3\dot{\Omega}_-^2) \right) \Theta(u - u_0) + \right. \\
& \left. + \ e^{\sqrt{3}\beta_+} N_+ + \frac{e^{-\sqrt{3}(\beta_++2\Omega_+)}}{N_+} (3\dot{\beta}_+^2 - 3\dot{\Omega}_+^2) \right) \Theta(u_0 - u) \right].
\end{aligned} \tag{4.68}$$

In order to find the Hamiltonian associated with this action, one first imposes the suitable boundary conditions, choosing the normalization factors as functions of t alone, i.e. $\chi_\pm(t)$. This is still consistent with the matching conditions on Σ and in fact, this choice extends their validity to the whole manifold. This implies that both regions of spacetime are no longer independent of each other. Aiming to incorporate this constraint into the action, a Lagrangian multiplier, $\Gamma(t)$, is introduced so to satisfy the Israel boundary conditions. Hence the action becomes:

$$\begin{aligned}
S' \ = \ & S + \int dt\, du\, \Gamma(t) \left(N_+ \chi_+ e^{\sqrt{3}\beta_+} - N_- \chi_- e^{\sqrt{3}\beta_-} \right) \\
= \ & \frac{1}{2} \int dt\, du \left[\left(e^{\sqrt{3}\beta_-} N_- + \frac{e^{-\sqrt{3}(\beta_-+2\Omega_-)}}{N_-} (3\dot{\beta}_-^2 - 3\dot{\Omega}_-^2) \right) \Theta(u - u_0) \right. \\
& \left. + \ e^{\sqrt{3}\beta_+} N_+ + \frac{e^{-\sqrt{3}(\beta_++2\Omega_+)}}{N_+} (3\dot{\beta}_+^2 - 3\dot{\Omega}_+^2) \right) \Theta(u_0 - u) \right] \\
& + \int dt\, du\, \Gamma(t) \left(N_+ \chi_+ e^{\sqrt{3}\beta_+} - N_- \chi_- e^{\sqrt{3}\beta_-} \right)
\end{aligned} \tag{4.69}$$

from which, following the standard procedure as can be seen in Refs. [105, 108], the following equations of motion are obtained:

$$\frac{\delta S'}{\delta N_-} = 0 \Leftrightarrow \frac{\delta S}{\delta N_-} - \Gamma(t)\chi_- e^{\sqrt{3}\beta_-} = 0, \tag{4.70a}$$

$$\frac{\delta S'}{\delta N_+} = 0 \Leftrightarrow \frac{\delta S}{\delta N_+} + \Gamma(t)\chi_+ e^{\sqrt{3}\beta_+} = 0, \tag{4.70b}$$

$$\frac{\delta S'}{\delta \Gamma(t)} = N_+\chi_+ e^{\sqrt{3}\beta_+} - N_-\chi_- e^{\sqrt{3}\beta_-} = 0. \tag{4.70c}$$

From these equations, one is able to recover the classical equations of motions for the KS spacetime on each side of the shell. Taking $u < u_0$, these equations yield:

$$e^{\sqrt{3}\beta_-}\Gamma(t)\chi_- = 0, \tag{4.71a}$$

$$e^{\sqrt{3}\beta_+} - 3\frac{e^{-\sqrt{3}(\beta_+ + 2\Omega_+)}}{N_+^2}(\dot{\beta}_+^2 - \dot{\Omega}_+^2) + e^{\sqrt{3}\beta_+}\Gamma(t)\chi_+ = 0, \tag{4.71b}$$

$$N_+\chi_+ e^{\sqrt{3}\beta_+} - N_-\chi_- e^{\sqrt{3}\beta_-} = 0. \tag{4.71c}$$

From the first of these equations one concludes that $\Gamma(t) = 0$ and the classical equation of motion for the inner spacetime is recovered, as well as the constraint in the third equation. Repeating this procedure for $u > u_0$ yields the classical equation of motion for the outer spacetime. Then, having established that the correct equations are obtained for the classical scenario, one is able to proceed to the quantization of the system. Solving Eqs. (4.70a) and (4.70b) for $\Gamma(t)$ and setting them equal to each other yields:

$$\frac{\delta S}{\delta N_+}\frac{1}{\chi_+ e^{\sqrt{3}\beta_+}} + \frac{\delta S}{\delta N_-}\frac{1}{\chi_- e^{\sqrt{3}\beta_-}} = 0. \tag{4.72}$$

Substitution of the action S from Eq. (4.68), results into:

$$\frac{1}{2\chi_+ e^{\sqrt{3}\beta_+}}\left(e^{\sqrt{3}\beta_+} - 3\frac{e^{-\sqrt{3}(\beta_+ + 2\Omega_+)}}{N_+^2}(\dot{\beta}_+^2 - \dot{\Omega}_+^2)\right)\Theta(u_0 - u)+$$
$$+ \frac{1}{2\chi_- e^{\sqrt{3}\beta_-}}\left(e^{\sqrt{3}\beta_-} - 3\frac{e^{-\sqrt{3}(\beta_- + 2\Omega_-)}}{N_-^2}(\dot{\beta}_-^2 - \dot{\Omega}_-^2)\right)\Theta(u - u_0) = 0, \tag{4.73}$$

then using Eq. (4.70c), it turns that:

$$\frac{1}{2\chi_+}\left(1 - 3\frac{\chi_+^2}{\chi_-^2}\frac{e^{-2\sqrt{3}(\beta_- + \Omega_+)}}{N_-^2}(\dot{\beta}_+^2 - \dot{\Omega}_+^2)\right)\Theta(u_0 - u)+$$
$$+ \frac{1}{2\chi_-}\left(1 - 3\frac{e^{-2\sqrt{3}(\beta_- + \Omega_-)}}{N_-^2}(\dot{\beta}_-^2 - \dot{\Omega}_-^2)\right)\Theta(u - u_0) = 0, \tag{4.74}$$

and the momenta associated with this Lagrangian are:

$$p_{\beta_-} = \frac{3e^{-\sqrt{3}(\beta_- + 2\Omega_-)}\dot{\beta}_-}{N_-}\Theta(u - u_0),$$

(4.75a)

$$p_{\beta_+} = \frac{3e^{-\sqrt{3}(\beta_- + 2\Omega_+)}\dot{\beta}_+}{N_-}\frac{\chi_+}{\chi_-}\Theta(-u + u_0),$$

(4.75b)

$$p_{\Omega_-} = -\frac{3e^{-\sqrt{3}(\beta_- + 2\Omega_-)}\dot{\Omega}_-}{N_-}\Theta(u - u_0),$$

(4.75c)

$$p_{\Omega_+} = -\frac{3e^{-\sqrt{3}(\beta_- + 2\Omega_+)}\dot{\Omega}_+}{N_-}\frac{\chi_+}{\chi_-}\Theta(-u + u_0),$$

(4.75d)

where Eq. (4.70c) has been used. Replacing the above momentum expressions into Eq. (4.74) yields:

$$e^{2\sqrt{3}\Omega_-}\left(p_{\beta_-}^2 - p_{\Omega_-}^2\right) + e^{2\sqrt{3}\Omega_+}\frac{\chi_-}{\chi_+}\left(p_{\beta_+}^2 - p_{\Omega_+}^2\right) - 3\left(\frac{\chi_-}{\chi_+}\Theta(u_0 - u) + \Theta(u - u_0)\right) = 0.$$

(4.76)

Therefore, this constraint can be identified with the Hamiltonian for the problem.

The most general solution for the problem does not imply any constraint on the arbitrary parameters χ_+ and χ_-. The Hamiltonian equation can indeed be rewritten as

$$H = \frac{1}{\chi_-}\left[e^{2\sqrt{3}\Omega_-}\left(p_{\beta_-}^2 - p_{\Omega_-}^2\right) - 3\Theta(u - u_0)\right] + \frac{1}{\chi_+}\left[e^{2\sqrt{3}\Omega_+}\left(p_{\beta_+}^2 - p_{\Omega_+}^2\right) - 3\Theta(u_0 - u)\right] = 0.$$

(4.77)

This Hamiltonian is similar to the one for the KS spacetime, but not quite the same. The difference lies in the matching conditions for the null boundaries which affect the manipulation of the constant term. Thus, the existence of a single Hamiltonian for the system is due to the stringent boundary conditions imposed by the Israel junction conditions. However, this Hamiltonian yields two independent constraints in the quantum mechanical problem.

Most importantly, the boundary condition constraint regulates the choice for the discontinuity of the lapse function across the shell. In the context of quantum gravity, the Hamiltonian constraint must be zero for every choice of lapse function. Here, in particular, this means that N_+ and N_- – and therefore χ_+ and χ_- – must be allowed to be arbitrary, and the constraint must vanish for every choice.

Before solving the Hamiltonian constraint, Eq. (4.77), in order to better understand the effect of the difference in the normalization factors, one considers the addition of the Eqs. (4.70a) and (4.70b) which yield:

$$e^{-\sqrt{3}\beta_-}\frac{\delta S}{\delta N_-} + e^{-\sqrt{3}\beta_+}\frac{\delta S}{\delta N_+} - \Gamma(t)\left(\chi_- - \chi_+\right) = 0.$$

(4.78)

Using the momentum definition, Eq. (4.75), this equation becomes:

$$e^{2\sqrt{3}\Omega_-}\left(p_{\beta_-}^2 - p_{\Omega_-}^2\right) + e^{2\sqrt{3}\Omega_+}\left(p_{\beta_+}^2 - p_{\Omega_+}^2\right) - 3 = -6\,\Gamma(t)\,(\chi_- - \chi_+). \qquad (4.79)$$

This expression behaves as a more general Hamiltonian for the system when the normalization factors are different. Obviously, this term vanishes when one considers the classical scenario, since, as seen previously, $\Gamma(t)$ vanishes on both sides of the shell. Nevertheless, this may not be the case in the quantized version. In order to check that this expression agrees with both equations of motion, one writes:

$$\left[e^{2\sqrt{3}\Omega_-}\left(p_{\beta_-}^2 - p_{\Omega_-}^2\right) + 6\,\Gamma(t)\chi_-\right] + \left[e^{2\sqrt{3}\Omega_+}\left(p_{\beta_+}^2 - p_{\Omega_+}^2\right) - 6\,\Gamma(t)\chi_+\right] - 3 = 0. \qquad (4.80)$$

In the next section, when the quantum solutions are calculated, one shall notice that the above equation corresponds to two different constraints, one multiplied by an arbitrary χ_+, holding on one side of the shell, and the other multiplied by an arbitrary χ_-, holding on the other side. This implies that, the lapse functions in the Hamiltonian formalism, N_+ and N_-, which are Lagrange multipliers, can be arbitrarily chosen as the constraint Eq. (4.71c), and it can thus be adjusted by the choice of the free parameters χ_+ and χ_-.

4.3 Quantization

4.3.1 Commutative scenario

For the quantum treatment of this problem, one will proceed to the quantization in three different ways. First, one uses Eq. (4.80) with canonical quantization in the minisuperspace in order to obtain the Wheeler-deWitt (WdW) equation for the system. This yields:

$$\left[e^{2\sqrt{3}\Omega_-}\left(\hat{p}_{\beta_-}^2 - \hat{p}_{\Omega_-}^2\right) + 6\,\Gamma\chi_- + e^{2\sqrt{3}\Omega_+}\left(\hat{p}_{\beta_+}^2 - \hat{p}_{\Omega_+}^2\right) - 6\,\Gamma\chi_+\right]\Psi\left(\beta_-,\Omega_-,\beta_+,\Omega_+\right) =$$
$$= +3\Psi\left(\beta_-,\Omega_-,\beta_+,\Omega_+\right). \qquad (4.81)$$

Using the Heisenberg-Weyl representation and ordering for the operators, one gets:

$$\left[e^{2\sqrt{3}\Omega_-}\left(\partial_{\beta_-}^2 - \partial_{\Omega_-}^2\right) - 6\,\Gamma\chi_- + e^{2\sqrt{3}\Omega_+}\left(\partial_{\beta_+}^2 - \partial_{\Omega_+}^2\right) + 6\,\Gamma\chi_+\right]\Psi\left(\beta_-,\Omega_-,\beta_+,\Omega_+\right) =$$
$$= -3\Psi\left(\beta_-,\Omega_-,\beta_+,\Omega_+\right). \qquad (4.82)$$

The above equation shows no crossing terms between quantities of inner and outer manifolds. This suggests that the wave function for the problem is separable and can be written as $\Psi(\beta_-,\Omega_-,\beta_+,\Omega_+) = \Psi_-(\beta_-,\Omega_-)\Psi_+(\beta_+,\Omega_+)$, which leads to the following decoupled equations:

$$\left[e^{2\sqrt{3}\Omega_-}\left(\partial_{\beta_-}^2 - \partial_{\Omega_-}^2\right) - 6\Gamma\chi_-\right]\Psi_-(\beta_-,\Omega_-) = \sigma^2\Psi_-(\beta_-,\Omega_-), \tag{4.83a}$$

$$\left[e^{2\sqrt{3}\Omega_+}\left(\partial_{\beta_+}^2 - \partial_{\Omega_+}^2\right) + 6\Gamma\chi_+\right]\Psi_+(\beta_+,\Omega_+) = -(\sigma^2+3)\Psi_+(\beta_+,\Omega_+), \tag{4.83b}$$

which can be written as:

$$e^{2\sqrt{3}\Omega_-}\left(\partial_{\beta_-}^2 - \partial_{\Omega_-}^2\right)\Psi_-(\beta_-,\Omega_-) = \lambda_-^2\Psi_-(\beta_-,\Omega_-), \tag{4.84a}$$

$$e^{2\sqrt{3}\Omega_+}\left(\partial_{\beta_+}^2 - \partial_{\Omega_+}^2\right)\Psi_+(\beta_+,\Omega_+) = -\lambda_+^2\Psi_+(\beta_+,\Omega_+), \tag{4.84b}$$

where $\lambda_-^2 = (\sigma^2 + 6\Gamma\chi_-)$ and $\lambda_+^2 = (\sigma^2 + 3 + 6\Gamma\chi_+)$.

Instead of quantizing Eq. (4.80), one can follow the same procedure for the equations of motion. This procedure yields the following equations:

$$e^{2\sqrt{3}\Omega_-}\left(\partial_{\beta_-}^2 - \partial_{\Omega_-}^2\right)\Psi_-(\beta_-,\Omega_-) = \bar{\lambda}_-^2\Psi_-(\beta_-,\Omega_-), \tag{4.85a}$$

$$e^{2\sqrt{3}\Omega_+}\left(\partial_{\beta_+}^2 - \partial_{\Omega_+}^2\right)\Psi_+(\beta_+,\Omega_+) = -\bar{\lambda}_+^2\Psi_+(\beta_+,\Omega_+), \tag{4.85b}$$

with $\bar{\lambda}_\pm = 3\left(\Gamma\chi_\pm - 1\right)$. From the above statements, it is possible to conclude that the equations obtained after quantization, using either method, are analogous, thus Eq. (4.80) is the right choice for Hamiltonian system, in the sense that it naturally yields the correct equations for the quantized system. One also notices that, although it is possible to write a single Hamiltonian for the system, it yields two separate constraints, as those resulted from the equations of motion, Eqs. (4.70). Additionally, one can quantize the system from Eq. (4.77), following a similar procedure (i.e. writing the equation as a sum of $+$ and $-$ contributions and equating each one to a constant), and the obtained equations would be the same.

The differential equations, Eq. (4.84), describe a dynamics similar to that one of the KS cosmological problem [92, 93], with real eigenvalues, $\lambda_\pm$, on each side of the shell. These auxiliary parameters, $\lambda_\pm$, are the quantum analogues of the Lagrange multiplier introduced in order to account for the constraint, Eq. (4.70c), and the different normalization factors on each side. The system can be described by a single Hamiltonian given that the Israel junction conditions, Eq. (4.70c), constrain the system. However, even with a single Hamiltonian, there are two conditions being imposed on the manifold, one for each side of the hypersurface. Therefore, the solution splits into two pieces, corresponding to the solutions given by Ψ_+ and Ψ_-:

$$\Psi_-(\beta_-,\Omega_-) = e^{i\sqrt{3}b_-\beta_-}K_{ib_-}\left(\frac{i\lambda_-}{\sqrt{3}}e^{-\sqrt{3}\Omega_-}\right), \tag{4.86a}$$

$$\Psi_+(\beta_+,\Omega_+) = e^{i\sqrt{3}b_+\beta_+} K_{ib_+}\left(\frac{\lambda_+}{\sqrt{3}}e^{-\sqrt{3}\Omega_+}\right),\tag{4.86b}$$

which results into the full solution:

$$\Psi(\beta_-,\Omega_-,\beta_+,\Omega_+) = e^{i\sqrt{3}b_-\beta_-} K_{ib_-}\left(\frac{i\lambda_-}{\sqrt{3}}e^{-\sqrt{3}\Omega_-}\right) e^{i\sqrt{3}b_+\beta_+} K_{ib_+}\left(\frac{\lambda_+}{\sqrt{3}}e^{-\sqrt{3}\Omega_+}\right).\tag{4.87}$$

The parameters $b_\pm$ can be regarded as momentum eigenvalues associated with the momentum operators, $\hat{p}_{\beta_\pm}$, since $\hat{p}_{\beta_\pm}\Psi_\pm = \sqrt{3}b_\pm\Psi_\pm$. This implies that $b_\pm \in \mathbb{R}$ and admits continuous values. Therefore, the eigenfunctions for this problem are a family of Bessel functions, labeled by the eigenvalue of the associated $\beta_\pm$ momentum operator. The squared modulus of the wave function is then written as

$$|\Psi|^2 = K_{ib_-}\left(\frac{i\lambda_-}{\sqrt{3}}e^{-\sqrt{3}\Omega_-}\right) K_{ib_-}^*\left(\frac{i\lambda_-}{\sqrt{3}}e^{-\sqrt{3}\Omega_-}\right) K_{ib_+}\left(\frac{\lambda_+}{\sqrt{3}}e^{-\sqrt{3}\Omega_+}\right) K_{ib_+}^*\left(\frac{\lambda_+}{\sqrt{3}}e^{-\sqrt{3}\Omega_+}\right),\tag{4.88}$$

which are depicted in Fig. 4.1. The plot for $|\Psi|^2$ is given in Fig. 4.2.

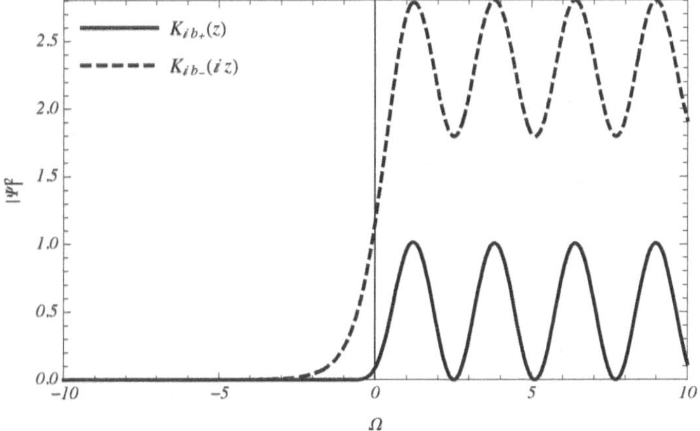

FIGURE 4.1: $|\Psi_-|^2$ and $|\Psi_+|^2$ as function of $\Omega_\pm$. The parameters are $b_\pm = 0.7$ and $\lambda_\pm = 2$.

It is relevant to point out that $|\Psi|^2$ does not depend on $\beta_\pm$, which can be interpreted as free wave parameters. However, according to the map, Eq. (4.4), $\beta_\pm$ are still relevant in distinguishing the point solutions from inner and outer spaces: i.e. any value for Ω is valid for both symmetrical points around the border between inner and outer spaces. Due to such a symmetrical behavior, for the physical limit given by $\Omega_\pm \to \infty$, one has the shell approaching either the singularity at $t = 0$ or the event horizon at $t = 2M$, which constrains the map to the interval $t \in [0, 2M]$. In particular, for $\Omega_\pm \to +\infty$, the wave function exhibits an oscillatory profile which can be interpreted as the absence of a meaningful limit for the

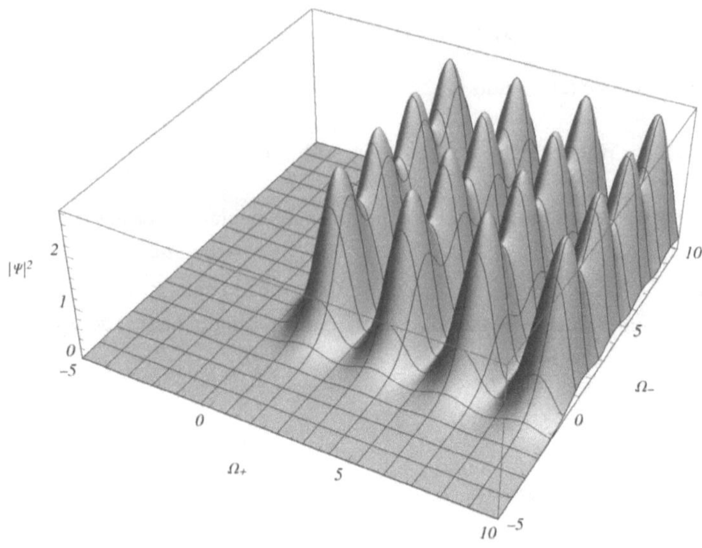

FIGURE 4.2: $|\Psi|^2$ in the coordinate space of $\Omega_\pm$. The parameters are $b_\pm = 0.7$ and $\lambda_\pm^2 = 2$.

quantization of the BH. Therefore no conclusions can be drawn from the limit corresponding to the BH singularity and the event horizon. The vanishing of the wave function for $\Omega_\pm <$ 0 is consistent with this function having a minimum and therefore to exclude arbitrarily negative values for $\Omega_\pm$ for a fixed BH mass.

4.3.2 Noncommutative scenario

One considers now a deformed Heisenberg-Weyl algebra, where position and momentum do not commute [26, 27, 92, 94]. The commutation relations are given by Eq. (1.19), adapted to the coordinates of the problem at hand:

$$[\hat{\beta}_{NC}, \hat{\Omega}_{NC}] = i\theta, \qquad [\hat{\beta}_{NC}, \hat{\pi}_\beta] = i, \qquad [\hat{\Omega}_{NC}, \hat{\pi}_\Omega] = i, \qquad [\hat{\pi}_\beta, \hat{\pi}_\Omega] = i\eta, \qquad (4.89)$$

where η and θ are new constants of Nature. These variables can be mapped into commutative ones, i.e. those supported by standard quantum mechanics (Heisenberg-Weyl algebra) relations, through a noncanonical transformation usually referred to as Darboux or Seiberg-Witten map [13]. In particular, a useful transformation is given by Eq. (1.22), which can be specialized to the BH problem in terms of the NC variables written as [26, 27, 92, 94]:

$$\hat{\beta}_{NC} = \hat{\beta} - \frac{\theta}{2}\hat{p}_\Omega, \quad \Omega_{NC} = \hat{\Omega} + \frac{\theta}{2}\hat{p}_\beta, \qquad (4.90)$$

for the position, and [26, 27, 92, 94]

$$\hat{\pi}_\beta = \hat{p}_\beta + \frac{\eta}{2}\hat{\Omega}, \quad \hat{\pi}_\Omega = \hat{p}_\Omega - \frac{\eta}{2}\hat{\beta},$$
(4.91)

for the momenta. With these transformations the Hamiltonian, Eq. (4.80), becomes:

$$\hat{H}_{NC} = e^{2\sqrt{3}(\hat{\Omega}_- + \frac{\theta}{2}\hat{p}_{\beta_-})}\left[\left(\hat{p}_{\beta_-} + \frac{\eta}{2}\hat{\Omega}_-\right)^2 - \left(\hat{p}_{\Omega_-} - \frac{\eta}{2}\hat{\beta}_-\right)^2\right] +$$
$$+ e^{2\sqrt{3}(\hat{\Omega}_+ + \frac{\theta}{2}\hat{p}_{\beta_+})}\left[\left(\hat{p}_{\beta_+} + \frac{\eta}{2}\hat{\Omega}_+\right)^2 - \left(\hat{p}_{\Omega_+} - \frac{\eta}{2}\hat{\beta}_+\right)^2\right] - \Gamma\left(\chi_- - \chi_+\right) - 3.$$
(4.92)

As was the case for the commutative problem, here the quantization either through the NC version of Eq. (4.80), Eq. (4.77) or through the equations of motion, Eqs. (4.70), yield the same differential equations for the wave function.

At this point, a few remarks are in order. Notice first that this Hamiltonian is symmetric with respect to the exchange of the inner and outer metric variables. Furthermore, each term of the Hamiltonian depends only on one of the sets of metric variables, since there are no crossed terms of "+" and "−" variables. The WdW equation for the problem is thus the following one:

$$\left\{e^{2\sqrt{3}(\hat{\Omega}_+ - i\frac{\theta}{2}\partial_{\beta_+})}\left[\left(-i\partial_{\beta_+} + \frac{\eta}{2}\hat{\Omega}_+\right)^2 - \left(-i\partial_{\Omega_+} - \frac{\eta}{2}\hat{\beta}_+\right)^2\right] + \Gamma\chi_+ + 3\right\}\Psi(\beta_-,\Omega_-,\beta_+,\Omega_+) +$$
$$+ \left\{e^{2\sqrt{3}(\hat{\Omega}_- - i\frac{\theta}{2}\partial_{\beta_-})}\left[\left(-i\partial_{\beta_-} + \frac{\eta}{2}\hat{\Omega}_-\right)^2 - \left(-i\partial_{\Omega_-} - \frac{\eta}{2}\hat{\beta}_-\right)^2\right] - \Gamma\chi_-\right\}\Psi(\beta_-,\Omega_-,\beta_+,\Omega_+) = 0,$$
(4.93)

for which likewise the commutative case, the same decoupling features for the wave function are found.

Thus, one can write the decoupled eigenvalue equations:

$$e^{2\sqrt{3}(\hat{\Omega}_- - i\frac{\theta}{2}\partial_{\beta_-})}\left[\left(-i\partial_{\beta_-} + \frac{\eta}{2}\hat{\Omega}_-\right)^2 - \left(-i\partial_{\Omega_-} - \frac{\eta}{2}\hat{\beta}_-\right)^2\right]\Psi_-(\beta_-,\Omega_-) = \lambda_-^2\Psi_-(\beta_-,\Omega_-),$$
(4.94a)
$$e^{2\sqrt{3}(\hat{\Omega}_+ - i\frac{\theta}{2}\partial_{\beta_+})}\left[\left(-i\partial_{\beta_+} + \frac{\eta}{2}\hat{\Omega}_+\right)^2 - \left(-i\partial_{\Omega_+} - \frac{\eta}{2}\hat{\beta}_+\right)^2\right]\Psi_+(\beta_+,\Omega_+) = -\lambda_+^2\Psi_+(\beta_+,\Omega_+),$$
(4.94b)

with $\lambda_\pm^2$ being the same as in the commutative case. In order to find the solution of the above equations, it is necessary to obtain a constant of motion associated with each of the separated Hamiltonians. It is straightforward to check that $\hat{B}_\pm = \hat{p}_{\beta_\pm} - \frac{\eta}{2}\hat{\Omega}_\pm$ commutes with $\hat{H}_\pm$ so that the solutions must obey the eigenvalue equations:

$$\hat{B}_\pm\Psi_\pm = \left(-i\partial_{\beta_\pm} - \frac{\eta}{2}\hat{\Omega}_\pm\right)\Psi_\pm = \sqrt{3}b_\pm\Psi_\pm,$$
(4.95)

and thus the wave function can be written as

$$\Psi_{\pm} = e^{\frac{i\beta_{\pm}}{2}(2\sqrt{3}b_{\pm}+\eta\Omega_{\pm})} F_{\pm}(\Omega_{\pm}). \tag{4.96}$$

Substituting the above solutions into Eqs. (4.94) leads to ordinary differential equations for the functions $F_{\pm}$:

$$F(\Omega)\left(3b_{\pm}^2 + \eta^2\Omega^2 + 2\sqrt{3}b_{\pm}\eta\Omega \pm e^{-\frac{\sqrt{3}}{2}(2\theta b_{\pm}+\eta\theta\Omega+4\Omega)}\lambda_{\pm}^2\right) + F''(\Omega) = 0. \tag{4.97}$$

Exact solutions for such equations are difficult to find for $\lambda_{\pm} \neq 0$. Otherwise, numerical solutions that satisfy the required boundary conditions are easily obtained. For the particular choice of $\lambda_{\pm} = 0$, an exact solution can be found for the ordinary differential equation:

$$F(\Omega_{\pm})(\sqrt{3}b_{\pm} + \eta\,\Omega_{\pm})^2 + F''(\Omega_{\pm}) = 0, \tag{4.98}$$

that is

$$F_{\pm}(\Omega_{\pm}) = c_1\, D_{-\frac{1}{2}}\left(\frac{(1+i)(\sqrt{3}b_{\pm} + \eta\,\Omega_{\pm})}{\sqrt{\eta}}\right) + c_2\, D_{-\frac{1}{2}}\left(-\frac{(1-i)(\sqrt{3}b_{\pm} + \eta\,\Omega_{\pm})}{\sqrt{\eta}}\right), \tag{4.99}$$

where $D_\nu(z)$ are parabolic cylinder functions, depicted in Fig. 4.3. They correspond to mirror reflected solutions with a damped oscillatory profile that vanish at $\Omega \to \pm\infty$. This feature arises from the introduction of the noncommutativity, more specifically, due to the momentum noncommutativity, since the parameter θ is absent from the WdW equation for $\lambda_{\pm} = 0$.

As previously argued, we should focus only on the behavior of the functions for $\Omega > 0$. One can see that, at least for $\lambda_{\pm} = 0$, the probability density in the limit $t \to 0$ vanishes. Furthermore, since the dependence of Ψ on $\beta_{\pm}$ appears only as a phase factor, this dependance disappears for $|\Psi|^2$. Thus, it is possible to conclude that the probability density in the limit $t \to 2M$ also vanishes. Since both geometries on either side of the shell have a vanishing probability density to reach the singularity, then, the shell, which is defined as the separation surface between these geometries, also has a vanishing probability density of collapsing into the singularity. The same arguments applies for $t = 2M$: the shell can never leave the event horizon.

Although the exact solution discussed above portrays the general characteristics of the solutions when noncommutativity is introduced, one must study the numerical solutions of Eq. (4.97) for $\lambda_{\pm} \neq 0$ since, for this cases, the effect of θ no longer disappears. By consistency, the boundary conditions to be used are the ones from the exact solutions for $\lambda_{\pm} = 0$, with $b_{\pm} = 0.5$ and $\eta = 0.7$. As expected, the choice of boundary conditions has no significative

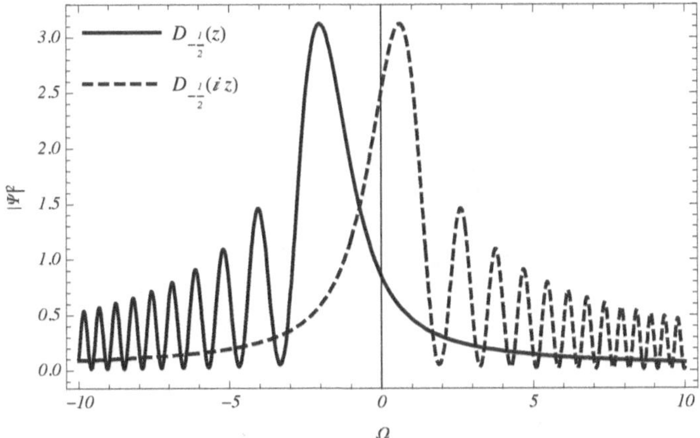

FIGURE 4.3: $|D_{-1/2}(z)|^2$ and $|D_{-1/2}(iz)|^2$ as functions of $\Omega_\pm$. The parameters are $\sqrt{3}b_\pm = 0.5$ and $\eta = 0.7$. Here $z = \frac{(1+i)(\sqrt{3}b_\pm + \eta\Omega)}{\sqrt{\eta}}$.

effect on the behavior of the wave functions, in particular, in the limit of interest, $\Omega \to \infty$. This can be better understood through the analysis of the asymptotic behavior of Eq. (4.97), which gives rise to:

$$F(\Omega_\pm)(\sqrt{3}b_\pm + \eta\,\Omega_\pm)^2 + F''(\Omega_\pm) = 0, \tag{4.100}$$

i.e. the same equation for the $\lambda_\pm = 0$ scenario. The effect of $\lambda_\pm$ is relevant only nearby $\Omega = 0$. The numerical results for some non-vanishing values of $\lambda_\pm$ and θ are shown in Fig. 4.4.

It can be seen that the damping behavior of the function found for the $\lambda_\pm = 0$ case is maintained. Furthermore, we find that damping is clearly a feature due to the momentum non-commutativity as it still remains for $\theta = 0$. Consequently, the conclusions drawn from the exact solution can be extended to the numerical ones. Namely, the vanishing of the probability density at the singularity and the event horizon at $t = 2M$. In order to compute the probability of the wave function, one must integrate over $\Omega_\pm$ on a surface of constant $\beta_\pm$. The integration measure is $\delta(\beta - \beta_c)\mathrm{d}\beta\,\mathrm{d}\Omega$ [92], which leads to:

$$P(\beta,\Omega) = \int \delta(\beta' - \beta_c)\mathrm{d}\beta'\,\mathrm{d}\Omega'\,|F(\Omega')|^2 \sim \int_0^{+\infty} \mathrm{d}\Omega'\,|F(\Omega')|^2. \tag{4.101}$$

Despite the vanishing of the wave function for $\Omega \to \infty$, the above integral is not convergent, so the obtained solution is not squared integrable. This is a feature also encountered for the NC black hole without a shell [27, 94], for the commutation relations Eqs. (4.89).

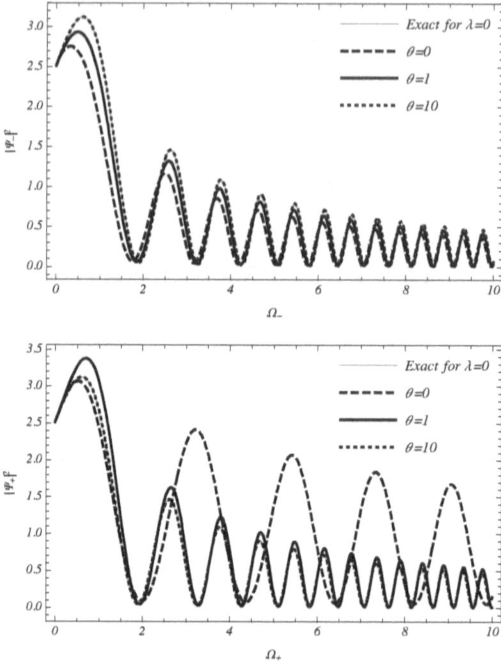

FIGURE 4.4: Numerical solutions of Eq. (4.97) for $\lambda_\pm = 1$ as function of Ω_- (first plot) and Ω_+ (second plot), with $b_\pm = 0.5$ and $\eta = 0.7$. Solutions are for $\theta = 0$ (dashed lines), $\theta = 1$ (solid lines), and $\theta = 10$ (dotted lines).

In any case, the effect of η alone leads to the damping of the wave function that is responsible for its regularization at the limit $\Omega \to \infty$. It is easy to see that the damping disapears for $\eta = 0$. Indeed considering Eq. (4.97) and in the limit $\eta \to 0$:

$$F(\Omega)\left(b_\pm^2 \mp e^{-\frac{1}{2}\sqrt{3}(2\theta b_\pm + 4\Omega)}\lambda_\pm\right) + F''(\Omega) = 0. \tag{4.102}$$

This equation admits exact solutions similar to the ones in Eq. (4.86), so that the complete wave functions $\Psi_\pm$ can be written as:

$$\Psi_-(\beta_-, \Omega_-) = e^{i\sqrt{3}b_- \beta_-} K_{ib_-}\left(\frac{\lambda_-}{\sqrt{3}} e^{-\frac{\sqrt{3}}{2}(\theta b_- + 2\Omega_-)}\right), \tag{4.103a}$$

$$\Psi_+(\beta_+, \Omega_+) = e^{i\sqrt{3}b_+ \beta_+} K_{ib_+}\left(\frac{i\lambda_+}{\sqrt{3}} e^{-\frac{\sqrt{3}}{2}(\theta b_+ + 2\Omega_+)}\right), \tag{4.103b}$$

which corresponds to a shift in the oscillatory behaviour of the wave function, Eqs. (4.86).

Chapter 5

Hořava-Lifshitz quantum cosmology

5.1 Introduction

Ideas about space-time provided by General Relativity (GR) are fundamentally linked to the classical dynamics of objects and particles. For arbitrarily small scales, say at Planck scale, spacetime can be considerably different. The transition between the ultimately quantum picture and the large scale properties of the Universe can depend on fundamental assumptions related to quantum cosmology and quantum gravity [108, 109]. Such issues concern the very nature of time itself and of quantum probabilities [110–114], as well as the cosmological boundary features and the initial singularity problem, and the ensued conditions for the onset of inflation [87, 115–120].

Aiming to set up a theory of quantum gravity, the canonical Hamiltonian formulation of the quantum cosmology driven by the Wheeler-DeWitt (WdW) equation for a wave function of the Universe [121] provides interesting scenarios to examine quantum effects in cosmology. However, the absence of a complete definition for the specific properties of quantum states resulting from this framework is in itself an obstacle for a detailed modeling of the primordial conditions of the Universe. It follows that this set up does not provide much room for observational implications.

The canonical WdW framework itself gives rise to various conceptual questions. One of them concerns the classification of infinite dimensional deformed algebras related to such a constraint equation, a discussion particularly relevant in loop quantum gravity [122]. On more physical grounds, the questions about the nature of time [110, 111, 114, 123, 124] stems from the conditions from which GR yields a constraint equation [87, 121], the WdW equation, and how it provides a clear-cut prescription for the time evolution.

In fact, classical cosmology does provide a robust insight into the understanding of the cosmic inventory and does allow for extracting, from the phenomenological data, quite relevant information about the existence of a dark sector that dominates the cosmic budget. Therefore, a procedure to fit the transition from quantum to classical descriptions of the cosmological framework is particularly relevant. In such a context, our proposal in this Chapter is to examine the classical cosmology arising from the Hořava-Lifshitz (HL) [125–128] quantum cosmological models [129–136]. Starting from quantum state solutions obtained from the HL WDW equation, an extension to the phase-space Wigner formalism for quantum mechanics (QM) is set up in order to provide a formulation to track the transition from quantum to classical descriptions.

As is well known, the derivation of testable predictions from quantum cosmology requires systematic approximations at semiclassical level and beyond, and most often, the minisuperspace approximation arises from symmetry arguments to reduce the number of degrees of freedom. Quantum cosmology in the context of minisuperspace models allows for the construction and investigation of the behavior of the wave function of the Universe obtained from the WdW equation. This framework deals with the relativistic space-time without a unique Hamiltonian to generate the dynamics. In any case, despite the fact that one has several choices for the definition of time and the ensued Hamiltonian evolution, these choices should lead to the same physics.

Since the considered symmetries are respected at classical level and that at quantum level they lead to a set of consistency conditions that restricts the possible choices of quantum cosmologies, in the HL cosmologies [125] to be considered in this Chapter, we shall be able to quantify some of the above-mentioned issues and to create a platform for describing quantum to classical transition scenarios.

Essentially, one assumes that the cosmological modifications due to QM can be computed from an extended version of the Liouville equation which describes the dynamical evolution of a quasi-Gaussian Wigner function in the Weyl-Wigner framework of QM. This framework depicts quite well the interplay between quantum and classical variables as it effectively takes place in constrained systems.

The key issue in the construction of the time-evolution of the quantum Wigner functions and of their associated probability currents is that it provides a quite illustrative picture of the quantum to classical transitions for the solutions of the WdW equation. In this context, the non-trivial configuration of initial quasi-Gaussian quantum states, as Hamiltonian eigenstate superpositions of projectable gravity scenarios according to HL cosmologies [125], can be investigated and their effects analytically quantified. For some simplified HL cosmological dynamics, the quantum to classical transition can be characterized by quantifiers of quantum distortion and nonclassicality, computed from Wigner currents.

Our analysis uses as a starting point the projectable HL gravity without detailed balance in a minisuperspace quantum cosmological Friedmann-Lemaître-Robertson-Walker (FLRW) universe without matter[1]. In fact, several cosmological scenarios have been investigated in the context of HL gravity [129–131], where projectable versions are those where the lapse function depends uniquely on time. This implies that the non-local classical Hamiltonian constraint of the GR must be integrated over spatial coordinates. This leads to modifications to the Friedman equation through an additional term that behaves like dust [132], although suppressed by integration [133]. Concomitantly, higher spatial curvature terms give rise to new cosmological features [134–136] involving, for instance, bouncing and oscillating solutions [137–139].

In what follows we shall consider the quantum cosmology of the HL quantum gravity in the minisuperspace approximation [131] and study its transition to classical cosmology. The onset of our study is the hyperbolic WdW equation [140, 141] and its solution for the HL gravity [131, 142], where the cosmological constant and matter components are included in the analysis [143].

It will be shown that in phase-space the quantum system is akin to the classical behavior, even though quantum properties are essential for a consistent description. In such a context, it is relevant to notice that the transition from quantum to classical behavior is often described with the phase-space Wigner formalism by decoherence processes [144–147]. Our analysis, however, provides results in terms of the Wigner currents and shows, through analytical expressions for the cosmological classical limit, an exact correspondence with the quantum cosmological description. The existence of such a quantum to classical correspondence is supported by the coincidence of classical trajectories with sharp peaks of the Wigner function, the most likely quantum region arising from a quasi-Gaussian expression obtained from the superposition of HL quantum states.

The outline of this Chapter is as follows. Section 5.2 is concerned with the HL quantum cosmologies in the minisuperspace approximation as reported in the literature. The WdW equation is re-obtained and a class of solutions which contain radiation, curvature and stiff matter contributions is worked out. In Section 5.3, an analytical expression for the Wigner function with quasi-Gaussian profile is obtained as a superposition of the HL solutions. It is pointed out that the parameters that drive the quantum superposition can be adjusted as to analytically fit the classical trajectories associated to the corresponding classical cosmology. The dynamics of such a quantum to classical transition can be discussed in terms of the Wigner flow analysis (cf. Appendices C and D) presented in Section 5.4, where quantum

---

[1] This model exhibits the main features of the HL gravity, in a completely covariant approach [126], which provides the detailed balance as a limiting hypothesis, being much simpler than non-projectable models [127, 128].

distortions are exactly obtained from defined Wigner currents. In particular, modifications due to the extension of the formalism to the description of models with bounces are also discussed. The inclusion of a cosmological constant component is performed in Section 5.5 in order to modify the Wigner currents which, in this case, are perturbatively re-obtained. A detailed analysis of the corresponding Wigner flow provides an expression for computing the age of the Universe in terms of the stiff matter contribution.

5.2 HL quantum cosmologies in the minisuperspace limit

The setup for our framework is the Einstein-Hilbert action given by

$$S = \frac{1}{16\pi G} \int d^4x \sqrt{-g}\, \mathcal{R}, \tag{5.1}$$

where $g = \det(g_{\mu\nu})$, $\mathcal{R} = R^{\mu\nu} g_{\mu\nu}$ is the scalar curvature, and $c = \hbar = 1$. The most general form of a $SO(4)$-invariant metric in a $M = \mathbb{R} \times S^3$ topology [88], in an homogeneous and isotropic space-time, is given by the line element of the Robertson-Walker (RW) metric,

$$ds^2 = -\sigma^2 \left[N(t)^2\, dt^2 - a(t)^2 \left(\frac{dr^2}{1 - g_c r^2} + r^2\, d\Omega^2 \right) \right], \tag{5.2}$$

where σ is a normalization constant, and $g_c = 0, +1$, and -1 denotes the curvature corresponding to $\mathbb{R}^3$, S^3 and H^3 hypersurfaces. In this case, one has $\sqrt{-g} = N(t)\, a(t)^3$, where the lapse function, $N(t)$, and the scale parameter, $a(t)$, are arbitrary non-vanishing functions of time, t, and $d\Omega^2 = d\theta^2 + \sin(\theta)^2 d\phi^2$.

More generically, in terms of metric components, the three-dimensional quantities used to describe the GR in the Arnowitt-Deser-Misner (ADM) formalism [148, 149] are g_{ij}, $\Pi_{ij} = \sqrt{-g}\, (\Gamma^0_{kl} - g_{kl}\, \Gamma^0_{mn}\, g^{mn})\, g^{ik} g^{jl}$, $N = (-g^{00})^{-1/2}$, and the shift vector, $N_i = g_{0i}$, where one has the connection, Γ^k_{ij}, as an independent quantity, with *latin* indices running from 1 to 3. Relevant to the discussion is the extrinsic curvature written as

$$K_{ij} = \frac{1}{2\sigma N} \left(-\frac{\partial g_{ij}}{\partial t} + \nabla_i N_j + \nabla_j N_i \right), \tag{5.3}$$

where ∇_i denotes the 3-dimensional covariant derivative, and for the metric Eq. (5.2),

$$K_{ij} = -\frac{1}{\sigma N} \frac{\dot{a}}{a} g_{ij}, \quad \text{with} \quad K = K^{ij} g_{ij} = -\frac{3}{\sigma N} \frac{\dot{a}}{a}, \tag{5.4}$$

as $N_i = 0$ (and $g_c = 1$). In this case, the 3-dim Ricci tensor components and the corresponding Ricci scalar are given, respectively, by $R_{ij} = 2g_{ij}/(\sigma^2 a^2)$ and $R = 6/\sigma^2 a^2$.

Finally, an interesting suggestion to achieve a renormalizable quantum gravity theory at high-energy is provided by the HL gravity, which at cosmological level is given by the action [133, 150],

$$
S_{HL} = \frac{M_{\mathrm{Pl}}^2}{2} \int d^3x\, dt\, N\sqrt{g} \left\{ K_{ij}K^{ij} - \lambda K^2 - g_0 M_{\mathrm{Pl}}^2 - g_1 R \right.
$$
$$
- g_2 M_{\mathrm{Pl}}^{-2} R^2 - g_3 M_{\mathrm{Pl}}^{-2} R_{ij}R^{ij} - g_4 M_{\mathrm{Pl}}^{-4} R^3 - g_5 M_{\mathrm{Pl}}^{-4} R \left(R^i{}_j R^j{}_i \right) \tag{5.5}
$$
$$
\left. - g_6 M_{\mathrm{Pl}}^{-4} R^i{}_j R^j{}_k R^k{}_i - g_7 M_{\mathrm{Pl}}^{-4} R \nabla^2 R - g_8 M_{\mathrm{Pl}}^{-4} \nabla_i R_{jk} \nabla^i R^{jk} \right\},
$$

where the balance of the curvature components is described by dimensionless coupling constants, g_i, with $i = 0, \ldots, 9$, and M_{Pl} denotes the Planck mass. Here it is worth to mention that the HL gravity is a framework in which the ultraviolet (UV) completion problem of GR [125] is circumvented by turning the gravity into a power-countable renormalizable theory at the UV fixed point. This is achieved by giving up the Lorentz symmetry at high-energies [125, 151], assuming that GR is recovered at an infra-red (IR) fixed point scenario of the HL gravity at low-energy scales. Such a Lorentz symmetry breaking is related to an anisotropic scaling of space and time, $r \to br$ and $t \to b^z t$, with b being a scale parameter and $z \neq 1$.

A sequence of assumptions carried out in Ref. [131] did reduce the number of degrees of freedom of the action Eq. (5.5). Firstly, one notices that the GR is recovered by setting $g_1 = -1$, through the rescaling of the time coordinate, and by taking the limit of $\lambda \to 1$, which recovers the full diffeomorphism invariance. However, λ must be a running constant, and there is no reason or symmetry that constrains it to the GR limit[2]. As for the cosmological constant, Λ, it can be written in Planck units as $\Lambda = g_0 M_{\mathrm{Pl}}^2/2$. Finally, isotropic and homogeneous conditions over g_{ij} impose constraints directly into the equations of motion, or through the substitution of the RW metric into the Lagrangian density[3]. Since the RW metric, Eq. (5.2), introduces an anisotropy between space and time, it does not affect the homogeneity. Then, it can be substituted into Eq. (5.5) and the integration over $d^3x \sqrt{\det g_{ij}}$ gives $2\pi^2$ so that the HL minisuperspace action [131] is re-written as

$$
S_{HL} = \frac{M_{\mathrm{Pl}}^2 \times 2\pi^2 \times 3(3\lambda - 1)\sigma^2}{2} \int dt N \left\{ \frac{-\dot{a}^2 a}{N^2} + \frac{6a}{3(3\lambda - 1)} - \frac{2\Lambda \sigma^2 a^3}{3(3\lambda - 1)} - \right.
$$
$$
\left. - M_{\mathrm{Pl}}^{-2} \times \frac{12}{3(3\lambda - 1)\sigma^2 a} \times (3g_2 + g_3) - M_{\mathrm{Pl}}^{-4} \times \frac{24}{3(3\lambda - 1)\sigma^4 a^3} \times (9g_4 + 3g_5 + g_6) \right\}.
$$
$$
\tag{5.6}
$$

---

[2]Even if the phenomenology suggests that λ is quite close to unity [143].

[3]In Ref. [?] the authors point to that such restrictions cannot be done over the Lagrangian unless one properly solves the arising constraints. Otherwise, the introduction of the RW metric does not lead, in general, to the same results obtained from the equations of motion [152].

By redefining the dimensionless constants [134]

$$g_C = \frac{2}{3\lambda - 1}, \tag{5.7}$$

$$g_\Lambda = \frac{\Lambda M_{\text{Pl}}^{-2}}{9\pi^2(3\lambda - 1)^2}, \tag{5.8}$$

$$g_R = 24\pi^2(3g_2 + g_3), \tag{5.9}$$

$$g_S = 288\pi^4(3\lambda - 1)(9g_4 + 3g_5 + g_6), \tag{5.10}$$

and choosing units simplified by the constraint $\sigma^2 \times 6\pi^2 \times (3\lambda - 1)M_{\text{Pl}}^2 = 1$, the minisuperspace action finally reads

$$S_{HL} = \frac{1}{2} \int dt \left(\frac{N}{a}\right) \left[-\left(\frac{a}{N}\dot{a}\right)^2 + g_C a^2 - g_\Lambda a^4 - g_R - \frac{g_S}{a^2} \right], \tag{5.11}$$

from which one identifies the canonical conjugate momentum associated to a as given by

$$\Pi_a = \frac{\partial \mathcal{L}}{\partial \dot{a}} = -\frac{a}{N}\dot{a}, \tag{5.12}$$

such that the HL minisuperspace Hamiltonian density becomes [131, 133]

$$H = \Pi_a \dot{a} - \mathcal{L} = \frac{1}{2}\frac{N}{a} \left(-\Pi_a^2 - g_C a^2 + g_\Lambda a^4 + g_R + \frac{g_S}{a^2} \right). \tag{5.13}$$

A discussion of the quantum mechanical problem resulting from the above Hamiltonian in the context of the WdW framework is presented in Refs. [125, 131, 133, 143, 151]. One notices that $g_C > 0$ stands for the curvature coupling constant and the sign of g_Λ follows the sign of the cosmological constant. In addition, both of them, g_C and g_Λ, are related to each other through a common degree of freedom, λ, such that the limit for the minisuperspace GR model is recovered by setting $\lambda = 1$. The sign of the coupling constants g_S and g_R, associated to stiff matter and radiation like contributions, respectively [133], does not affect the stability of the HL gravity [125, 134].

By following the canonical quantization strategy [87, 121], the canonical conjugate momentum is promoted to an operator [87],

$$\Pi_a \mapsto -i\frac{d}{da} \quad \text{such that} \quad \Pi_a^2 = -\frac{1}{a^q}\frac{d}{da}\left(a^q\frac{d}{da}\right),$$

where the choice of q does not affect the semiclassical analysis [153]. The classical minisuperspace Hamiltonian is thus promoted to an operator which acts on the wave function of the Universe $\bar{\psi}(a)$ such that the final form of the WdW equation (for $q = 0$) is then given by

$$\frac{1}{2} \left(\frac{d^2}{da^2} - g_c a^2 + g_\Lambda a^4 + g_R + \frac{g_s}{a^2} \right) \bar{\psi}(a) = 0, \tag{5.14}$$

from which one identifies the quantum potential given by

$$V(a) = \frac{1}{2} \left(g_c a^2 - g_\Lambda a^4 - g_R - \frac{g_s}{a^2} \right). \tag{5.15}$$

This is an one-dimensional Schrödinger-like equation constrained by a vanishing eigenvalue, $E = 0$, for which a complete analysis of its eigenvalues has already been performed in Refs. [131, 134]. However, the constraint, $E = 0$, in a certain sense, blurs the understanding of the *meaning of time* in the above procedure. Given that the classical time evolution is set by the Friedmann equation, any discussion of quantum to classical transition should take into account the conditions for the matching between the quantum and the classical frameworks.

In quantum cosmology, the notion of time is established following different premises since, in GR, time is observer-dependent and not absolute, and time translations are not generated by an observable, such as the energy, but by an expression which, on physical states, is constrained to vanish. Time translations are part of the transformations allowed by the algebra of the fundamental space-time symmetries [109, 122, 154]. Obtaining a consistent time definition stems from the evolution that a system experiences after imposing some smearing conditions into the canonical Hamiltonian formalism [112, 155, 156]. The simplest way to establish a time evolution, for specific matter/field contributions, is through the so called deparameterization procedure: one simply picks the variable that classically depends monotonically on time and interpret it as time itself [109, 154, 157]. On more general grounds, the absence of a covariant treatment leads to different quantum theories and one cannot find a unitary transformation to consistently identify how the observables vary [158–163].

Given these difficulties, it is sensible to expect that the choice of the *time parameter* should not affect the physical results. The point in this Chapter, which shall be postponed to the discussion of the quantum to classical transitions at the end of Section III, is that the identification of a parametric canonical variable, τ, with the quantum mechanical time, implicitly identified by a Hamiltonian correspondence with $i\,\partial/\partial\tau$, should be a suitable map for the *classical* results so to establish a natural bridge between quantum to classical cosmological descriptions. The association of the time parameter with the radiation energy has a clearly classical appeal since, classically, the correspondence between time and inverse square temperature of the cosmic background radiation is consistent with the phenomenological analysis that accounts for the cosmic energy density inventory. Although lacking a covariant framework, the association between time and temperature is consistent with suggestions for the origin of time asymmetry according to which the arrow of time does not reverse at an eventual contraction of the Universe [164, 165]. A consistent definition for such an extra degree of freedom, τ, should constrain the properties of the Hamiltonian operator. Herein, the notion of an extra degree of freedom associated to an environment set by radiation, with an associated unitary operator, $H_\nu \equiv i\partial/\partial\tau$, for the coordinate τ, allows one to rewrite the wave function as $\tilde{\psi}(a) \equiv \tilde{\psi}(a, \tau)$ as to have

$$H_\nu \tilde{\psi}(a, \tau) = E\tilde{\psi}(a, \tau), \quad \Rightarrow \quad \tilde{\psi}(a, \tau) = \psi(a) \exp(-i\, E\, \tau), \tag{5.16}$$

with E and τ in Planck units (cf. after Eqs. (5.6)-(5.10)). It is important to point out that, in our approach, the correspondence between time and radiation energy provides the elements for composing a large set of energy eigenstate quantum superpositions that exhibit a Schrödinger-like time evolution. In fact, the identification of the time variable in quantum cosmology is associated to the transition from quantum to classical dynamics [166] where several competing frameworks have been considered, either for pure states or for open systems. Our assumption resembles operationally the effects of unimodular quantum cosmology [167] where time arises from a secondary constraint which, once integrated, leads to an expression similar to that of Eq. (5.16). In that case, the associated energy, E, is actually the cosmological constant, and the time parameter is extrinsically identified with the Hamiltonian operator. As in our approach, the time is not an internal dynamical unconstrained phase-space variable. Otherwise, besides the usual WdW Hamiltonian constraint written in the form of Eq. (5.16), in our approach there is no additional secondary constraint as considered in the unimodular framework. Such an interpretation for τ is not different from the inclusion of some canonical coordinate related to additional matter/field/radiation components into the action, with the attribute of a time variable [157, 168, 169].

In fact, the parameter g_R can be rewritten as $g_R = g_\gamma + g_\nu = g_\gamma + 2E$, in order to give a physical meaning to the eigenvalue E, where the parameters g_γ and g_ν are, for instance, identified with photon and with massless neutrino contributions, respectively.

By introducing the elements of the above discussion and setting $a = g_c^{-1/4}x$, one obtains the re-parameterized equation

$$\left\{ \frac{d^2}{dx^2} - x^2 - \frac{4\alpha^2 - 1}{4x^2} + 2(\alpha + 2n + 1) + \ell x^4 \right\} \psi_n^\alpha(a(x)) = 0, \tag{5.17}$$

with $\ell = g_c^{-3/2} g_\Lambda$, and where one identifies the parameters α and n related to $E \rightarrow E_n = g_c^{1/2}(2n + 1)$, and to the coupling constants by $g_R = 2\alpha\, g_c^{1/2}$ and $g_s = -(4\alpha^2 - 1)/4$, for the wave function given by

$$\psi_n^\alpha(a(x)) = g_c^{\frac{1}{8}} \varphi_n^\alpha(x). \tag{5.18}$$

An exact solution of the above quantum mechanical problem can be obtained for $g_\Lambda = 0$. The solution is also valid for $0 \lesssim a \ll 1$ for $g_\Lambda \neq 0$ [131], however this case demands for a perturbative treatment. For $g_\Lambda = 0$, one has

$$\varphi_n^\alpha(x) = 2^{1/2}\,\Theta(x)\,N_n(\alpha)\,x^{\alpha+\frac{1}{2}}\,\exp(-x^2/2)\,L_n^\alpha(x^2), \tag{5.19}$$

where L_n^α are the *associated Laguerre polynomials*, $\Theta(x)$ is the *step-unity function* that constrains the result to $x > 0$, and $N_n(\alpha)$ is the normalization constant given by

$$N_n(\alpha) = \left(\frac{n!}{\Gamma(n + \alpha + 1)}\right)^{1/2}, \tag{5.20}$$

where $\Gamma(s)$ is the *gamma function*. The quantum mechanical potential, as depicted in Fig. 5.1,

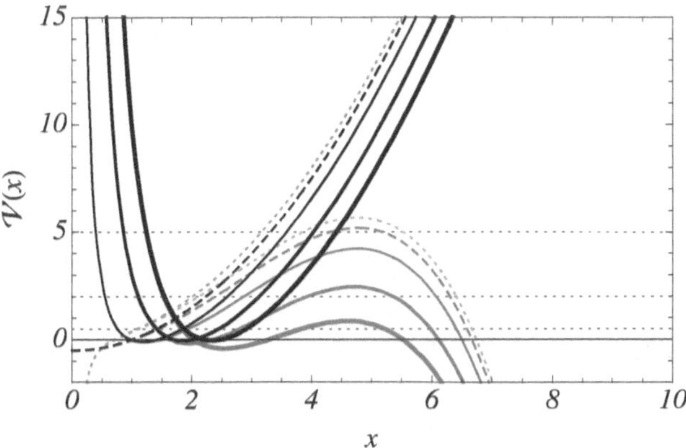

FIGURE 5.1: Re-parameterized potentials $g_C^{-1/2} V(a(x)) \equiv \mathcal{V}(x)$ as function of the cosmo-logical parameter $x = g_C^{1/4} a$ for the cosmological constant associated parameter $\ell = 0$ (black lines) and $\ell = 0.022$ (red lines). The behavior of the red line plots do not qualitatively change for $\ell \neq 0$. The plots are for $\alpha = 0$ (dotted lines), $1/2$ (dashed lines), $3/2$ (thinest solid lines), $7/2$ and $11/2$ (thickest solid lines). For $\alpha = 0$ one has $g_S < 0$ and $g_R = 0$ which leads to an undesirable singularity, and for $\alpha = 1/2$ one has no contribution from stiff matter, $g_S = 0$, which corresponds to the one-dimensional harmonic oscillator limit in the Schrödinger equation. Horizontal dashed lines refer to the classical energy correspondence for $E = 1/2$, 2 and 5.

includes the contribution from the $g_\Lambda \neq 0$ term, and it is now written as

$$g_C^{-1/2} V(a(x)) \equiv \mathcal{V}(x) = \frac{1}{2}\left[x^2 + \frac{4\alpha^2 - 1}{4x^2} - 2\alpha - \ell x^4\right]. \tag{5.21}$$

Solution Eq. (5.19) can be investigated in the framework of phase-space Wigner formalism in order to provide the elements that can be compared with the classical dynamics driven by $V(a) = g_C^{1/2}\,\mathcal{V}(x)$.

5.3 Wigner function for a quasi-Gaussian superposition and the classical limit

The Wigner function and the Weyl transform establish an alternative framework to connect quantum observables to expectation values as it provides a suitable insight into the quantum behavior and its classical limit. For a generic quantum state identified by the wave function, $f_\alpha(a)$, the Wigner function can be read as the Fourier transform of the off-diagonal terms of the associated density matrix, $f_\alpha^*(a + y_a) f_\alpha(a - y_a)$, which is given by the so-called Weyl transform,

$$W^\alpha(a, \Pi_a) = \frac{1}{\pi} \int_{-\infty}^{+\infty} dy_a \, \exp(2i\,\Pi_a\,y_a)\, f_\alpha^*(a + y_a)\, f_\alpha(a - y_a). \tag{5.22}$$

This is interpreted as a real-valued quasi-probability distribution, as $W^\alpha(a, \Pi_a)$ can, in principle, assume local negative values. This formulation has the operational advantage of exhibiting all the information content of the state vector in the phase-space, where operators, $\hat{a}$ and $\hat{\Pi}_a$, have been converted into c-numbers.

A re-dimensionalized form of Eq. (5.22) is written as

$$W^\alpha(a, \Pi_a) \equiv W^\alpha(x, p) = \frac{1}{\pi} \int_{-\infty}^{+\infty} dy \, \exp(2i\,p\,y)\, F_\alpha^*(x + y)\, F_\alpha(x - y), \tag{5.23}$$

where one has identified: $a = g_c^{-1/4}x$, $y_a = g_c^{-1/4}y$, $\Pi_a = g_c^{1/4}p$, and $f_\alpha(a(x)) = g_c^{1/8}F_\alpha(x)$, such that

$$\int_{-\infty}^{+\infty} da \int_{-\infty}^{+\infty} d\Pi_a \, W^\alpha(a, \Pi_a) = \int_{-\infty}^{+\infty} dx \int_{-\infty}^{+\infty} dp \, W^\alpha(x, p) = 1. \tag{5.24}$$

Generically speaking, to connect quantum operators to averaged observables, the trace of the product of two operators, $\hat{O}_1$ and $\hat{O}_2$, is given by the phase-space integral of the product of their Weyl transform [28, 170],

$$Tr_{\{x,p\}}\left[\hat{O}_1\hat{O}_2\right] = \iint dx\, dp\, O_1^W(x, p)\, O_2^W(x, p), \tag{5.25}$$

which, in terms of the properties of the density matrix operator, $\hat{\rho}$, gives

$$Tr_{\{x,p\}}\left[\hat{\rho}\hat{O}\right] = \langle O \rangle = \iint dx\, dp\, W(x, p)\, O^W(x, p), \tag{5.26}$$

which establishes the natural generalization of the Wigner function, from pure states to statistical mixtures, with the purity $Tr[\hat{\rho}^2]$ given by

$$Tr_{\{x,p\}}[\hat{\rho}^2] = 2\pi \iint dx\, dp\, W(x, p)^2, \tag{5.27}$$

where the factor 2π is introduced in order to satisfy the constraints: $Tr[\hat{\rho}^2] = Tr[\hat{\rho}] = 1$, for pure states.

Getting back to the HL solutions with the assumed decoupling between space-like coordinate, a, and time, τ, as set by Eq. (5.16), a time dependent version of $F_\alpha(x) \sim F_\alpha(x, \tau)$ is given by a superposition of the eigenstates from Eq. (5.19) as

$$F_\alpha(x, \tau) = \mathcal{N}_F \sum_{n=0}^{\infty} c_n^\alpha(\tau)\, \varphi_n^\alpha(x), \tag{5.28}$$

where $\mathcal{N}_F$ is a normalization constant. Once identifying the parameter $g_c^{1/2}$ with an energy quantity, $\omega/2$, given in Planck units, T_{Pl}^{-1}, where T_{Pl} is the Planck time, as to have $E_n = \omega\,(n + 1/2)$, one can construct a quasi-Gaussian superposition if $c_n^\alpha \equiv u^n\, N_n^{-1}(\alpha)\, \exp(-(i/2)\,\omega\tau)$, with $u \equiv u(\beta, \tau) = \exp(-\beta + i\,\omega\,\tau)$, where $\omega\tau$ is a dimensionless quantity, and β is an arbitrary weight parameter which constrains the expansion coefficients to $|u| < 1$. By substituting $\varphi_n^\alpha(x)$ into the above superposition, one obtains the complex function

$$\begin{aligned}
F_\alpha(x, \tau) &= \mathcal{N}_F\, \Theta(x)\, x^{\alpha + \frac{1}{2}}\, \exp(-x^2/2) \sum_{n=0}^{\infty} u^n\, L_n^\alpha(x^2) \\
&= \mathcal{N}_F\, \Theta(x)\, x^{\alpha + \frac{1}{2}}\, (1 - u)^{-(1+\alpha)} \exp\left[-\frac{1}{2}\left(\frac{1+u}{1-u}\right) x^2 \right]
\end{aligned} \tag{5.29}$$

with

$$\mathcal{N}_F = \left[\frac{(1 - e^{-2\beta})^{1+\alpha}}{2\Gamma(1 + \alpha)} \right]^{1/2}.$$

Then the normalized probability distribution associated to F_α is explicitly written in terms of τ and β as

$$|F_\alpha(x, \tau)|^2 = \frac{2\,\mu_{(\beta,\tau)}^{1+\alpha}}{\Gamma(1+\alpha)}\, \Theta(x)\, x^{1+2\alpha}\, \exp(-\mu_{(\beta,\tau)} x^2), \tag{5.30}$$

with

$$\mu_{(\beta,\tau)} = \frac{\sinh(\beta)}{\cosh(\beta) - \cos(\omega\tau)}. \tag{5.31}$$

Such an auxiliary function, $\mu_{(\beta,\tau)}$, helps one to define a comoving coordinate, $\tilde{x}_{(\beta,\tau)}$ implicitly given by

$$\tilde{x}_{(\beta,\tau)}^2 = \frac{\alpha + 1}{\langle F_\alpha(x, \tau)|x^2|F_\alpha(x, \tau)\rangle}\, x^2 = \mu_{(\beta,\tau)} x^2, \tag{5.32}$$

which defines the comoving profile of an infinitesimal element of the quasi-Gaussian probability, $dx\,|F_\alpha(x, \tau)|^2 = d\tilde{x}\,|F_\alpha(\tilde{x})|^2$, with

$$|F_\alpha(\tilde{x})| = \frac{2}{\Gamma(1+\alpha)}\, \Theta(\tilde{x})\, \tilde{x}^{1+2\alpha}\, \exp(-\tilde{x}^2), \tag{5.33}$$

whose behavior is depicted in the first plot of Fig. 5.2 for several values of α.

From the above expression, one sees that it is easy to mathematically compose a Gaussian state from a superposition of F_α functions in α, which, however, might bring some unphysical features since different α values mix different stiff matter profiles. Nevertheless, the relevant point is that the obtained quasi-Gaussian profile reproduces the classical trajectories in the phase-space.

Through the calculation of the Wigner function, one can get $W_\alpha(x, p)$ for (semi-)integer values of α. The less interesting case, of course, corresponds to $\alpha = 1/2$ for which one has $g_s = 0$ and the quantum mechanical potential is reduced to the harmonic oscillator one. For $\alpha = 0$ one has $g_s > 0$, with $g_\gamma = 0$. Solutions for this case (for which the quantum mechanical potential is exhibited by Fig. 5.1) are not physically meaningful given the presence of a singularity at the origin, which compromises the evolution of the quantum eigenstates in terms of τ to an asymptotic squeezed state around $a = 0$.

By substituting the expression from Eq. (5.29) into Eq. (5.23) one obtains

$$
\begin{aligned}
W^\alpha(x, p; \tau) \;=\;& \frac{2\,\mu_{(\beta,\tau)}^{1+\alpha}}{\pi\,\Gamma(1+\alpha)} \int_{-\infty}^{+\infty} dy\, \Theta(x+y)\Theta(x-y)\,(x^2-y^2)^{\frac{1}{2}+\alpha} \\
& \exp\left(-\mu_{(\beta,\tau)}(x^2+y^2)\right)\,\exp\left(2\,i\,y(p+\tilde\mu_{(\beta,\tau)}\,x)\right) \quad\quad (5.34) \\[2mm]
\;=\;& \frac{2\,\mu_{(\beta,\tau)}^{1+\alpha}}{\pi\,\Gamma(1+\alpha)}\, x^{2(1+\alpha)}\,\exp\left(-\mu_{(\beta,\tau)}x^2\right) \\
& \int_{-1}^{+1} ds\,(1-s^2)^{\frac{1}{2}+\alpha}\,\exp\left(-\mu_{(\beta,\tau)}\,x^2\,s^2\right)\,\exp\left(2\,i\,x\,s(p+\tilde\mu_{(\beta,\tau)}\,x)\right),
\end{aligned}
$$

where the dependence on τ has been included through $\mu_{(\beta,\tau)}$ (cf. Eq. (5.31)) and

$$
\tilde\mu_{(\beta,\tau)} = -\frac{\sin(\omega\tau)}{\cosh(\beta)-\cos(\omega\tau)}. \quad\quad (5.35)
$$

In Eq. (5.34), the integrating variable has been simplified to $y \sim x\,s$, which is helpful for verifying the normalization conditions for $W^\alpha(x, p; \tau)$ (cf. the calculations performed in the Appendix I). By observing the function parity over the symmetric limits, the integration can be further simplified by introducing the power series expansion,

$$
\cos(2z) = \sum_{k=0}^{\infty}(-1)^k\,\frac{2^{2k}}{(2k)!}\,z^{2k}, \qu\quad (5.36)
$$

and using

$$
\begin{aligned}
2\int_0^{+1} ds\,(1-s^2)^{\frac{1}{2}+\alpha}\,s^{2k}\,\exp\left(-\mu_{(\beta,\tau)}\,x^2\,s^2\right) =& \\
\Gamma(3/2+\alpha)\Gamma(1/2+k)\,{}_1\mathcal{F}_1(1/2+k, 2+\alpha+k, -\mu_{(\beta,\tau)}\,x^2\,s^2),& \quad (5.37)
\end{aligned}
$$

where $_1\mathcal{F}_1$ is the *confluent hypergeometric function of the first kind*. One thus obtains the final form of the Wigner function as

$$W^\alpha(x, p; \tau) = \frac{2}{\sqrt{\pi}} \frac{\Gamma(3/2 + \alpha)}{\Gamma(1 + \alpha)} (\mu_{(\beta,\tau)} x^2)^{1+\alpha} \exp\left(-\mu_{(\beta,\tau)} x^2\right) \tag{5.38}$$
$$\sum_{k=0}^{\infty} \frac{(-1)^k (p\,x + \tilde{\mu}_{(\beta,\tau)} x^2)^{2k}}{\Gamma(1+k)\Gamma(2+k+\alpha)} \, _1\mathcal{F}_1(1/2 + k, 2 + \alpha + k, -\mu_{(\beta,\tau)} x^2 s^2),$$

which is not very helpful given that it involves an infinite sum over k.

However, for semi-integer values of α, written as $\alpha = 1/2 + v$, with $v = 0, 1, 2, \ldots$, one has the finite sum

$$(1 - s^2)^{\frac{1}{2}+\alpha} = (1 - s^2)^{1+v} = \sum_{k=0}^{1+v} (-1)^k \frac{s^{2k}(1+v)!}{k!(1+v-k)!} = \sum_{k=0}^{1+v} \frac{\Gamma(3/2 + \alpha)}{\Gamma(3/2 + \alpha - k)\Gamma(1 + k)}, \tag{5.39}$$

which, after substitution into Eq. (5.34), leads to an expression that is easier to handle,

$$W^\alpha(x, p; \tau) = \frac{4}{\pi} \frac{\Gamma(3/2 + \alpha)}{\Gamma(1 + \alpha)} (\mu\, x^2)^{1+\alpha} \exp\left(-\mu\, x^2\right) \sum_{k=0}^{\frac{1}{2}+\alpha} \frac{(-1)^k}{\Gamma(1+k)\Gamma(3/2 + k + \alpha)}$$
$$\int_0^{+1} ds\, s^{2k} \exp\left(-\mu\, x^2 s^2\right) \cos\left(2\,x\,s(p + \tilde{\mu}\,x)\right)$$
$$= \frac{1}{\sqrt{\pi}} \frac{\Gamma(3/2 + \alpha)}{\Gamma(1 + \alpha)} (\mu\, x^2)^{1+\alpha} \exp\left(-\mu\, x^2\right) \sum_{k=0}^{\frac{1}{2}+\alpha} \frac{x^{-(1+2k)}}{\Gamma(1+k)\Gamma(3/2 + k + \alpha)}$$
$$\frac{d^k}{d\mu^k}\left[\mu^{-1/2} \exp\left[-\frac{(p + \tilde{\mu}x)^2}{\mu}\right] \left(\mathrm{Erf}[\zeta(\mu, \tilde{\mu})] + h.c.\right)\right], \tag{5.40}$$

with $\zeta(\mu, \tilde{\mu}) = \mu^{1/2}(x + i\,\mu^{-1}(p + \tilde{\mu}x))$, and where the subindex $_{(\beta,\tau)}$ has been suppressed [4].

---

[4]To go further, given that the *Error function*, $\mathrm{Erf}[\ldots]$, has analytically well-defined derivatives written in terms of *Hermite polynomials*, $\mathcal{H}_{k-1}(z)$,

$$\frac{\sqrt{\pi}}{2} \frac{d^k}{dz^k} \mathrm{Erf}(z) = (-1)^{k-1} \mathcal{H}_{k-1}(z) \exp(-z^2), \tag{5.41}$$

one can write

$$\mathrm{Erf}\left[\mu^{1/2}\left(x + i\frac{p + \tilde{\mu}x}{\mu}\right)\right] + h.c = \frac{4}{\sqrt{\pi}} \exp\left[\frac{(p + \tilde{\mu}x)^2}{\mu}\right] \sum_{k=0}^{\infty} \mathcal{H}_{k-1}\left[i\,\mu^{-1/2}(p + \tilde{\mu}x)\right] \frac{(\mu\, x^2)^{k+\frac{1}{2}}}{(2k + 1)!}, \tag{5.42}$$

which leads to

$$W^\alpha(x, p; \tau) = \frac{4}{\pi} \frac{\Gamma(3/2 + \alpha)}{\Gamma(1 + \alpha)} (\mu\, x^2)^{1+\alpha} \exp\left(-\mu\, x^2\right)$$
$$\sum_{k=0}^{\frac{1}{2}+\alpha} \frac{x^{-2k}}{\Gamma(1+k)\Gamma(3/2 + k + \alpha)} \frac{d^k}{d\mu^k} \sum_{q=0}^{\infty} \mathcal{H}_{q-1}\left[i\,\mu^{-1/2}(p + \tilde{\mu}x)\right] \frac{(\mu\, x^2)^q}{(2q + 1)!}. \tag{5.43}$$

5.3.1 Coincident classical trajectories

Considering the phase-space element $d\Pi_a \, da \equiv dp \, dx$ and the associated Poisson brackets given by $\{a, \Pi_a\}_{PB} \equiv \{x, p\}_{PB} = 1$, the dynamics of the Hamiltonian associated to Eq. (5.17), for $\ell = 0$, is given by the corresponding Hamilton equations for x and p assuming that the time dependence is driven by $H = H_C - E$, with

$$H(x,p) = \frac{g_C^{1/2}}{2}\left[p^2 + x^2 + \frac{4\alpha^2 - 1}{4x^2} - 2\alpha \right]. \tag{5.44}$$

The classical trajectories are characterized by the condition $H_C = E$, and

$$p_C = \pm\left[2\Delta - \left(x_C^2 + \frac{4\alpha^2 - 1}{4x_C^2} \right) \right]^{1/2}, \qquad \text{with} \quad \Delta = \alpha + g_C^{-1/2}E, \tag{5.45}$$

where E is the classical energy and the index "C" has been introduced to denote classical quantities. The evaluation of the Poisson brackets yields

$$\dot{p}_C = \{p_C, H\}_{PB} = -g_C^{1/2}\left(x_C^2 + \frac{4\alpha^2 - 1}{4x_C^2} \right), \tag{5.46}$$

$$\dot{x}_C = \{x_C, H\}_{PB} = +g_C^{1/2} p_C, \tag{5.47}$$

where "*dots*" denote derivatives with respect to the classical time. Through the constraint Eqs. (5.45) and (5.47) one has

$$\dot{x}_C = \pm g_C^{1/2}\left[2\Delta - \left(x^2 + \frac{4\alpha^2 - 1}{4x^2} \right) \right]^{1/2}, \tag{5.48}$$

which can be rewritten in terms of $\eta = x_C^2$ as

$$\dot{\eta} = \pm g_C^{1/2}\left[\frac{1 - 4\alpha^2}{4} + 2\eta\Delta - \eta^2 \right]^{1/2}. \tag{5.49}$$

By identifying the time variable with τ from the quantum framework, classical and quantum dynamics can be compared. Solving Eq. (5.49), one obtains

$$\eta_\mp(\tau) \equiv \eta_\mp(\tau) = \Delta \mp \varkappa^{1/2}\sin(\vartheta + \omega\tau), \tag{5.50}$$

with ϑ arbitrary and

$$\varkappa = \frac{1}{4} + 2g_C^{-1/2}\alpha E + g_C^{-1}E^2.$$

For $\vartheta = \pi/2$ one has the classical solution

$$x_{C\mp}^2(\tau) = \Delta \mp \varkappa^{1/2}\cos(\omega\tau), \tag{5.51}$$

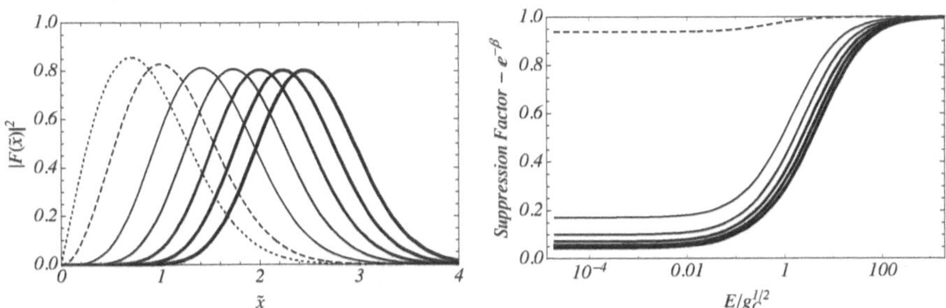

FIGURE 5.2: (First plot) Comoving modulus of the quasi-Gaussian wave function, $F_\alpha(\tilde{x})$, for the comoving coordinate $\tilde{x}$. (Second plot) Suppression factor, $\exp(-\beta)$, for the energy parameter $E/g_C^{1/2}$. The plots are for $\alpha = 0$ (dotted lines), $1/2$ (dashed lines), $3/2$ (thinest solid lines), $5/2, 7/2, 9/2$ and $11/2$ (thickest solid lines).

which constrains the arguments of the Wigner function, $W^\alpha(x, p; \tau)$, to time independent values, i.e. the *stationary profile* of $W^\alpha(x_c(\tau), p_c(\tau); \tau)$ along the classical trajectory guarantees that $W^\alpha(x, p; \tau)$ returns time-dependent averaged values of the quantum observables, x and p, that match the classical results, likewise to what happens with a Gaussian function for the harmonic oscillator problem. This can be verified by observing the dynamical behavior of the arguments of $W^\alpha(x_c(\tau), p_c(\tau); \tau)$,

$$\tilde{x}^2_{(\beta,\tau)} = \mu_{(\beta,\tau)} x_c^2(\tau) = \frac{\sinh(\beta)\left(\Delta - \varkappa^{1/2}\cos(\omega\tau)\right)}{\cosh(\beta) - \cos(\omega\tau)} \equiv \left(\alpha^2 - \frac{1}{4}\right)^{1/2}, \tag{5.52}$$

where μ is given by Eq. (5.31) and one constrains the β parameter to

$$\beta = \operatorname{arctanh}\left[\frac{(4\alpha^2 - 1)^{1/2}}{2(\alpha + g_C^{-1/2}E)}\right], \tag{5.53}$$

which is depicted in the second plot of Fig. 5.2. From Eqs.(5.31) and (5.35) one also notices that $\dot{\mu} = \tilde{\mu}\mu\omega$ and therefore[5]

$$\frac{d}{d\tau}\left(x\,p + \tilde{\mu}\,x^2\right) = (\mu\omega)^{-1}\frac{d}{d\tau}\left(\mu\,x^2\right) = 0. \tag{5.54}$$

Eq.(5.54) gives the exact constraint on the quantum mechanical superposition that corresponds to the quantum state $W^\alpha(x_c(\tau), p_c(\tau); \tau)$, which, in its turn, reproduces the classical profile of the evolution of the scale parameter, a: the classical cosmological solutions can be traced back to its quantum origin.

---

[5]In fact, after simple manipulations one shows that $x\,p + \tilde{\mu}\,x^2 = 0$ for the above choice of parameters.

5.4 HL Wigner flow and phase-space local quantum distortions

For the phase-space described in terms of the coordinates (a, Π_a), the conservation of probabilities is translated into the continuity equation (cf. Eq. (D.1) from Appendix II),

$$
\begin{aligned}
\frac{\partial W^\alpha}{\partial t} + \frac{\partial J_a^\alpha}{\partial a} + \frac{\partial J_{\Pi_a}^\alpha}{\partial \Pi_a}
&= \frac{\partial W^\alpha}{\partial t} + 2g_c^{1/2}\left[\frac{\partial}{\partial x}\left(\frac{J_a^\alpha}{2g_c^{1/4}}\right) + \frac{\partial}{\partial p}\left(\frac{J_{\Pi_a}^\alpha}{2g_c^{3/4}}\right)\right] \\
&= \frac{1}{\omega}\frac{\partial W^\alpha}{\partial \tau} + \frac{\partial J_x^\alpha}{\partial x} + \frac{\partial J_p^\alpha}{\partial p} \\
&= 0,
\end{aligned}
\tag{5.55}
$$

where again it has been set that $t \equiv \tau$, with $2g_c^{1/2} = \omega$, in order to recast the Wigner current
components as

$$
\begin{aligned}
J_x^\alpha(x, p; \tau) &= (2g_c^{1/4})^{-1} J_a^\alpha = (2g_c^{1/4})^{-1}\,\Pi_a\, W^\alpha(a, \Pi_a; \tau) = \frac{p}{2} W^\alpha(x, p; \tau), \tag{5.56} \\
J_p^\alpha(x, p; \tau) &= (2g_c^{3/4})^{-1} J_{\Pi_a}^\alpha \\
&= -(2g_c^{3/4})^{-1} \sum_{k=0}^{\infty} \left(\frac{i}{2}\right)^{2k} \frac{1}{(2k+1)!}\left[\left(\frac{\partial}{\partial a}\right)^{2k+1} V(a)\right]\left(\frac{\partial}{\partial \Pi_a}\right)^{2k} W^\alpha(a, \Pi_a; \tau) \\
&= -\frac{1}{2}\sum_{k=0}^{\infty}\left(\frac{i}{2}\right)^{2k}\frac{1}{(2k+1)!}\left[\left(\frac{\partial}{\partial x}\right)^{2k+1}\mathcal{V}(x)\right]\left(\frac{\partial}{\partial p}\right)^{2k} W^\alpha(x, p; \tau). \tag{5.57}
\end{aligned}
$$

where $\mathcal{V}(x)$ is given by Eq. (5.21), and the overall multiplying factor (an one-half factor in
the final form for this case) is not relevant and can be absorbed by the normalization definitions of the parameter ω. The above phase-space vectors follow the fluid equations from
Appendix II (for a canonical coordinates obeying $[\xi_x, \xi_p] = i\hbar$ converted into $[a, \Pi_a] = i$)
through a generalization of the Wigner flow formalism to the WdW quantum cosmological
approach.

Notice that by truncating the above sum at $k = 0$ one recovers the classical results. Quantum
and non-linear corrections arise from the contributions due to the infinite expansion,

$$
-\frac{1}{2}\sum_{k=1}^{\infty}\left(\frac{i}{2}\right)^{2k}\frac{1}{(2k+1)!}\left[\left(\frac{\partial}{\partial x}\right)^{2k+1}\mathcal{V}(x)\right]\left(\frac{\partial}{\partial p}\right)^{2k} W^\alpha(x, p; \tau),
\tag{5.58}
$$

which, for $\mathcal{V}(x)$, can be accurately described through an analytical expression. The constant
and the harmonic oscillator contribution from $\mathcal{V}(x)$ are easy to manipulate and do not introduce any kind of quantum back-flow [171, 172]. Since one has the potential proportional to
x^2, the harmonic oscillator Wigner current contribution is simply given by

$$
J_p^{\alpha(HO)}(x, p; \tau) = -\frac{x}{2} W^\alpha(x, p; \tau),
\tag{5.59}
$$

i.e. the amplitude for classical and quantum cases are the same.

Otherwise, one should pay some attention to the contribution due to the potential term which is proportional to $1/x^2$ (cf. Eq. (5.21)) and, eventually, to perturbative contributions due to ℓx^4 (cf. Sec. V)).

As to perform the analytical calculations, one first notices that

$$\left(\frac{\partial}{\partial x}\right)^{2k+1}\frac{1}{x^2} = -(2k+2)\frac{(2k+1)!}{x^{2k+3}},\qquad(5.60)$$

and

$$\left(\frac{\partial}{\partial p}\right)^{2k}W^\alpha(x,\,p;\,\tau) = \frac{1}{\pi}\int_{-\infty}^{+\infty}dy\,(2\,i\,y)^{2k}\,\exp(2\,i\,p\,y)\,F_\alpha^*(x+y)\,F_\alpha(x-y).\qquad(5.61)$$

One can then work out the sum in Eq. (5.58) related to the term proportional to $1/x^2$ in $V(x)$ as to have

$$-\frac{1}{2}\sum_{k=1}^{\infty}\left(\frac{i}{2}\right)^{2k}\frac{1}{(2k+1)!}\left[\left(\frac{\partial}{\partial x}\right)^{2k+1}\frac{1}{x^2}\right]\left(\frac{\partial}{\partial p}\right)^{2k}W^\alpha(x,\,p;\,\tau) =$$

$$= \frac{1}{x^3}\int_{-\infty}^{+\infty}dy\,\sum_{k=1}^{\infty}(-1)^{2k}\frac{(2k+1)!}{(2k+1)!}(k+1)\left(\frac{y}{x}\right)^{2k}\exp(2\,i\,p\,y)\,F_\alpha^*(x+y)\,F_\alpha(x-y)$$

$$= \frac{1}{\pi x^3}\int_{-\infty}^{+\infty}dy\,\frac{d}{d\varepsilon}\left(\sum_{k=1}^{\infty}\varepsilon^{k+1}\right)\exp(2\,i\,p\,y)\,F_\alpha^*(x+y)\,F_\alpha(x-y)$$

$$= \frac{x}{\pi}\int_{-\infty}^{+\infty}dy\,(x^2-y^2)^{-2}\,\exp(2\,i\,p\,y)\,F_\alpha^*(x+y)\,F_\alpha(x-y)$$

$$= x\mu^2\frac{\Gamma(\alpha-1)}{\Gamma(\alpha+1)}W^{\alpha-2}(x,\,p;\,\tau),\qquad(5.62)$$

where $\varepsilon = y^2/x^2 < 1$ has been considered, and without loss of generality, due to the Heaviside distribution in each function F, since $y \in [-x, x]$ then

$$\frac{d}{d\varepsilon}\left(\sum_{k=1}^{\infty}\varepsilon^{k+1}\right) = (1-\varepsilon)^{-2}.$$

The contribution from Eq. (5.62) must be multiplied by $(1-4\alpha^2)/8$ as to match the contribution from $V(x)$ which, once it is added to the contribution from Eq. (5.59), results into

$$J_p^\alpha(x,\,p;\,\tau)\;=\;-\frac{x}{2}\left(W^\alpha(x,\,p;\,\tau)+\frac{1-4\alpha^2}{4}\mu^2\frac{\Gamma(\alpha-1)}{\Gamma(\alpha+1)}W^{\alpha-2}(x,\,p;\,\tau)\right),\qquad(5.63)$$

from which one obtains

$$J_p^{\alpha(Cl)}(x,\,p;\,\tau)\;=\;-\frac{1}{2}W^\alpha(x,\,p;\,\tau)\left(x+\frac{1-4\alpha^2}{4x^3}\right),\qquad(5.64)$$

for the classical limit.

In fact, for $\tau = 0$, quantum fluctuations are highly suppressed by the quasi-Gaussian profile of the Wigner function, as it has been identified by the light-dark color scheme in Fig. (5.3). The Wigner flow stagnation points are defined by orange-green crossing lines, where $J_x^\alpha = J_p^\alpha = 0$, so that quantum features are completely suppressed for the series expansion, Eq. (5.58), truncated at $k = 0$. The classical trajectory portrait is shown as a collection of black dashed lines. The non-Liouvillian behavior [173] is depicted in Fig. (5.4) for arbitrary choices of α.

Thus, by comparing expressions for the classical and quantum Wigner currents, one is able to quantify the quantum fluctuations due to $\Delta J_p^\alpha(x, p; \tau) = J_p^\alpha(x, p; \tau) - J_p^{\alpha(Cl)}(x, p; \tau)$ over any specific volume $\Delta p\, \Delta x$ of the phase-space.

Classicality

Given that $W^\alpha(x, p; \tau)$ corresponds to a pure state, the global fluid behavior sets vanishing values for the rate of change of purity [174–177], $\dot{\mathcal{P}}$ (see the Appendix I). Such a behavior is interpreted as due to phase space symmetry and closure properties [174–177]. Therefore, only local quantum fluctuations can be identified and quantified as a measure of non-classicality for the Wigner flow [178].

For a periodic motion defined by closed trajectories as, for instance, those obtained from Eqs. (5.47)-(5.49), one can associate the classical trajectory to the two-dimensional boundary surface from the Wigner flow.

According to the results discussed in Appendix II [178], the local features of non-classicality for periodic motions defined by classical trajectories (cf. Eqs. (5.47)-(5.49)) can be quantified in terms of an integrated periodic probability flux enclosed by the classical surface, $\mathcal{C}$, given by (cf. Eqs. (D.13)-(D.16))

$$\frac{D}{D\tau}\mathrm{Prob}_{(\mathcal{C})}\bigg|_{\tau=\frac{2\pi}{\omega}} = -\int_0^{\frac{2\pi}{\omega}} d\tau\, \Delta J_p^\alpha(x_c(\tau),\, p_c(\tau); \tau)\, p_c(\tau). \tag{5.65}$$

From Eqs. (5.63) and (5.64), one writes

$$\Delta J_p^\alpha(x_c(\tau),\, p_c(\tau); \tau) = \frac{1 - 4\alpha^2}{8}\left(x_c(\tau)\,\mu^2\,\frac{\Gamma(\alpha - 1)}{\Gamma(\alpha + 1)}W^{\alpha-2}(x_c(\tau),\, p_c(\tau);\, \tau)\right.$$
$$\left. + x_c^{-3}(\tau)\, W^\alpha(x_c(\tau),\, p_c(\tau);\, \tau)\right), \tag{5.66}$$

with $x_c^2(\tau)$ given by Eq. (5.51), and $p_c^2(\tau)$ obtained from Eq. (5.45). In fact, according to the results from Eqs. (5.45)-(5.51) the above integral can be evaluated by noticing the stationary

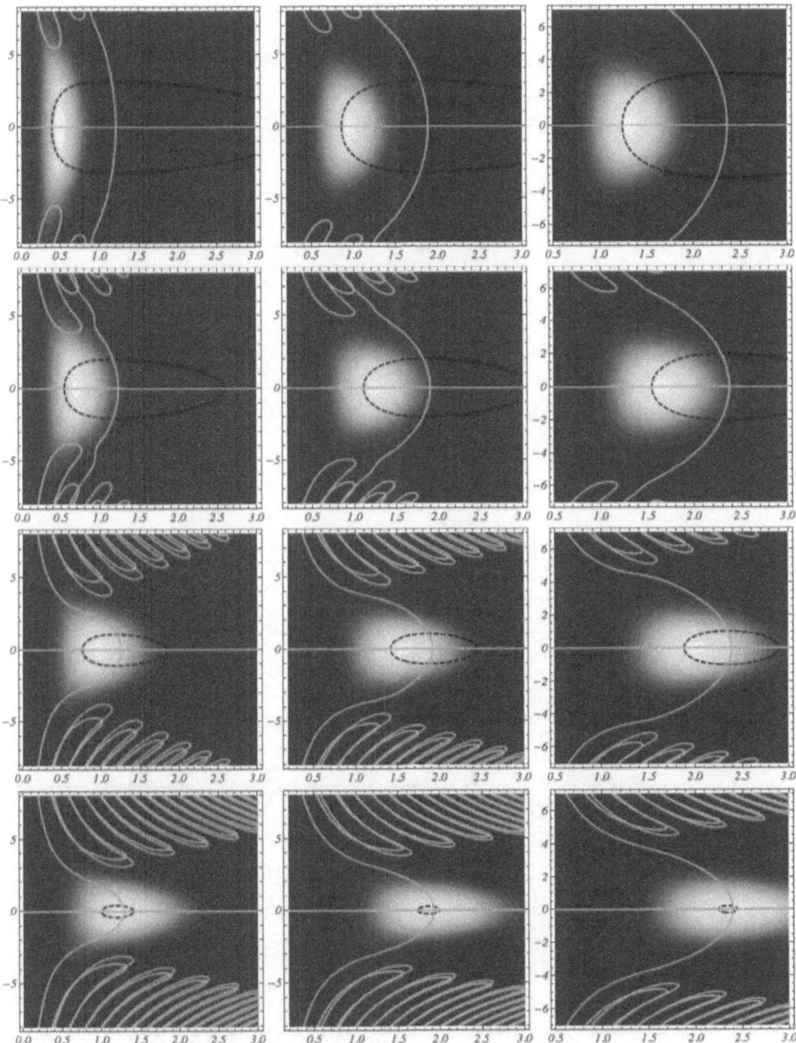

FIGURE 5.3: (Color online) Features of the Wigner flow for the quasi-Gaussian wave function, $W^\alpha(x, p; \tau)$, in the $x - p$ plane, at $\tau = 0$. Green contour lines are for $J_x^\alpha(x, p; 0) = 0$ and orange contour lines are for $J_p^\alpha(x, p; 0) = 0$. Green (orange) contour lines are bounds for the reversal of the Wigner flow in the $x(p)$ direction. The plots are for $E = 5, 2, 0.5$, and 0, from top to bottom, and for $\alpha = 3/2, 7/2$, and $11/2$, from left to right. The color scheme background shows the Wigner function profiles of W^α for $\omega\tau = 0$, with the details of the domains quantum fluctuations bounded by green and orange lines.

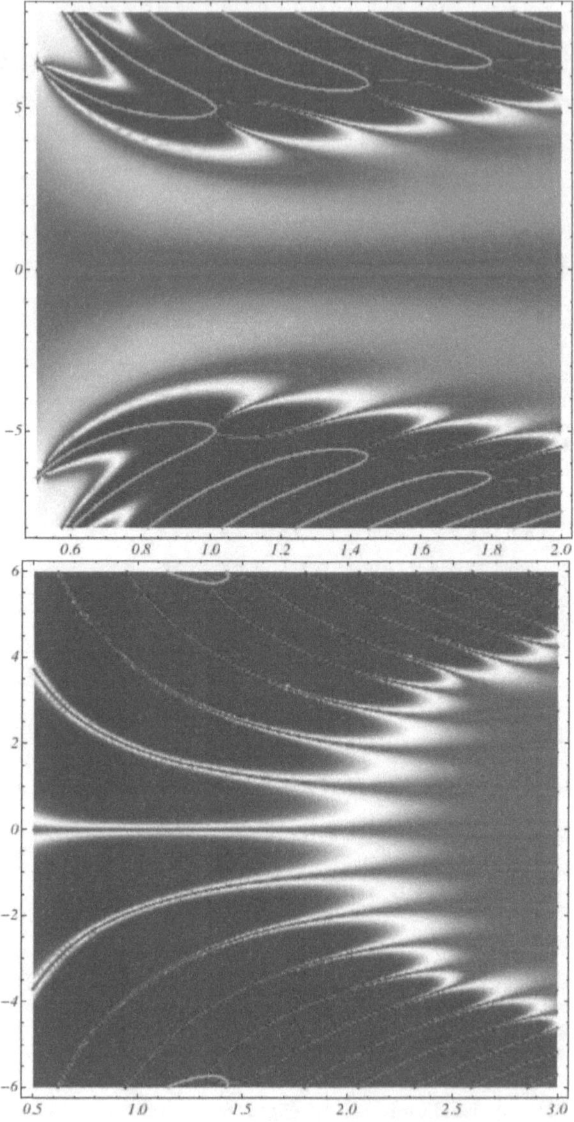

FIGURE 5.4: (Color online) Behavior of the Liouvillian quantifier parameterized by $\mathrm{sech}(\boldsymbol{\nabla}_{\xi} \cdot \mathbf{u})$ for the quasi-Gaussian Wigner function, $W^{\alpha}(x, p; \tau)$ in the phase-space ($x - p$ plane). Red-lines are for $\boldsymbol{\nabla}_{\xi} \cdot \mathbf{u} = 0$ and the *TemperatureMap* color scheme (from blue-regions, $\mathrm{sech}(\boldsymbol{\nabla}_{\xi} \cdot \mathbf{u}) \sim 0$, to red-regions, $\mathrm{sech}(\boldsymbol{\nabla}_{\xi} \cdot \mathbf{u}) \sim 1$) reinforces the approximated Liouvillian behavior for red-regions. Green-lines mark the zeros of the coincident values of $W^{\alpha}(x, p; 0) = J_x^{\alpha}(x, p; 0)$, where the $\boldsymbol{\nabla}_{\xi} \cdot \mathbf{u}$ becomes unbounded (maximal non-Liouvillian behavior). The plots are for $\alpha = 3/2$ (top) and $11/2$ (bottom).

behavior of the multiplying factor

$$S^\alpha \equiv S^\alpha(x_c(\tau), p_c(\tau); \tau) = \frac{1-4\alpha^2}{8}\left(x^4(\tau)\,\mu^2\,\frac{\Gamma(\alpha-1)}{\Gamma(\alpha+1)}W^{\alpha-2}(x_c(\tau), p_c(\tau); \tau)\right.$$

$$\left. +W^\alpha(x_c(\tau), p_c(\tau); \tau)\right), \qquad (5.67)$$

with $dS^\alpha/d\tau = 0$, which sets

$$\frac{D}{D\tau}\mathrm{Prob}_{(C)}\Big|_{\tau=\frac{2\pi}{\omega}} = -S^\alpha \int_0^{\frac{2\pi}{\omega}} d\tau \, p_c(\tau)\, x_c^{-3}(\tau)$$

$$= -S^\alpha \int_0^{\frac{2\pi}{\omega}} d\tau \, \varkappa^{1/2} \sin(\omega\tau)\left(\Delta - \varkappa^{1/2}\cos(\omega\tau)\right)^{-4}$$

$$= \frac{3}{\omega}S^\alpha \int_0^{\frac{2\pi}{\omega}} d\tau \left(\Delta - \varkappa^{1/2}\cos(\omega\tau)\right)^{-3}\Big|_{\tau=0}^{\tau=\frac{2\pi}{\omega}}$$

$$= 0. \qquad (5.68)$$

Given that, due to the quantum fluctuations, the integrand is non-vanishing along the parametric classical trajectory, the above result is quite auspicious as it depicts the quantum to classical transition of the HL scenario here described.

A second approach to quantify the quantum fluctuations can be established by the averaged value of $\Delta J_p^\alpha(x, p; \tau)$ over all the phase-space volume,

$$\Delta^\alpha(\tau) = \int_0^\infty dx \int_{-\infty}^{+\infty} dp \, W^\alpha \, \Delta J_p^\alpha(x, p; \tau) =$$

$$\frac{4\alpha^2-1}{2}\frac{\mu^{2(1+\alpha)}}{\pi^2\Gamma^2(1+\alpha)}\int_0^\infty dx\, x^{1+4\alpha}\exp\left(-2\mu x^2\right)\int_{-\infty}^{+\infty} dp \exp\left(2\,i\,x\,p\,(s+r)\right)\times$$

$$\int_{-1}^{+1}ds\int_{-1}^{+1}dr\,[(1-r^2)(1-s^2)]^{\frac{1}{2}+\alpha}\left[(1-s^2)^{-2}-1\right]\times$$

$$\exp\left(-\mu x^2\,(r^2+s^2)\right)\exp\left(2i\,\tilde{\mu}_{(\beta,\tau)}x^2\,(s+r)\right). \qquad (5.69)$$

By following the strategy suggested by Eqs. (C.6)-(C.8) from Appendix I, one obtains

$$\Delta^\alpha(\tau) = \frac{4\alpha^2-1}{2}\frac{\mu^{2(1+\alpha)}}{\pi\,\Gamma^2(1+\alpha)}\int_0^\infty dx\, x^{4\alpha}\exp\left(-2\mu x^2\,(1+s^2)\right)\times$$

$$\int_{-1}^{+1}ds\,(1-s^2)^{1+2\alpha}\left[(1-s^2)^{-2}-1\right] \qquad (5.70)$$

$$= \mu^{\frac{3}{2}}\frac{4\alpha^2-1}{2^{\frac{5}{2}+2\alpha}\pi}\frac{\Gamma(1/2+2\alpha)}{\Gamma^2(1+\alpha)}\int_{-1}^{+1}ds\,(1+s^2)^{\frac{1}{2}+2\alpha}(1-s^2)^{1+2\alpha}\left[(1-s^2)^{-2}-1\right],$$

where the periodic time dependence is factorized from the α dependent parameters.

The behavior of this quantity is shown in Fig. 5.5, where one notices that the influence of

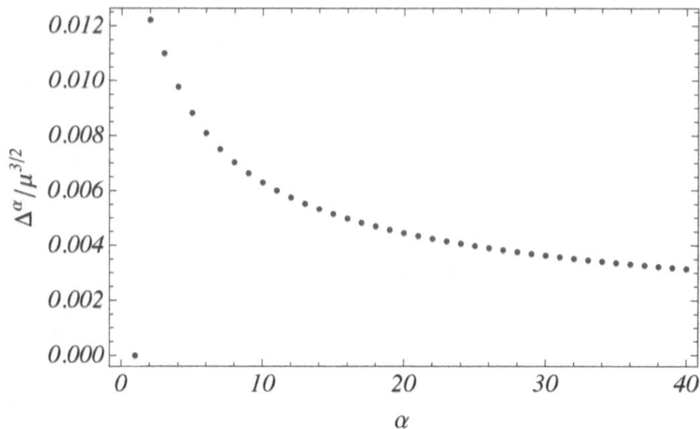

FIGURE 5.5: Quantum back-reaction quantifier, $\Delta^\alpha(\tau)$, normalized by the time dependence on $\mu_{(\beta,\tau)}^{\frac{3}{2}}$, as a function of α. Notice that for $\alpha = 1/2$ one recovers the (classical) harmonic oscillator result.

quantum fluctuations are suppressed by increasing values of α, as it was qualitatively depicted in Fig. 5.3. The action from Eq. (5.11) indeed yields an increasing value proportional to α as the t integration is performed. This is consistent with the expectation that classicality corresponds to the action, Eq. (5.11), $S \gg 1$ (in Planck units). The results from Fig. 5.5 are complemented by the flux map of Figs. 5.6 and 5.7.

Turning back to the qualitative analysis provided by the integrated flux, one should notice that the evolution of the Wigner function fits accurately the classical trajectory, as predicted by the sequence of Eqs. (5.30)-(5.54) from Sec. III.

Once one has $\alpha - 2 \geq 1/2$ as to satisfy the condition for obtaining the analytical expression for $J_p^\alpha(x, p; \tau)$, the qualitative interpretation of the results from Figs. 5.6 and 5.7 is not affected by the choice of E and α. Quantitatively, the quantum distortion is suppressed by increasing values of α, which corresponds to the transition of quantum trajectories into classical ones.

Notice that the quantum superposition described by $W^\alpha(x, p; \tau)$ follows the classical trajectory (thick black dashed-lines) in spite of the presence of quantum back-reaction effects (residual red arrows) due to $W^{\alpha-2}(x, p; \tau)$ along the p direction.

The Wigner flow stagnation points, which are typical quantum features, are defined by orange-green crossing lines, with $J_x^\alpha = J_p^\alpha = 0$. In the quantum cosmology context, the stagnation points vanish for $M_{\text{Pl}}^{-1} \to 0$. They are identified by clockwise and anti-clockwise vortices (winding number equals to $+1$ and -1), separatrix intersections and saddle flows (winding number equals to 0). As a compensating effect, the contra-flux fringes (delimited

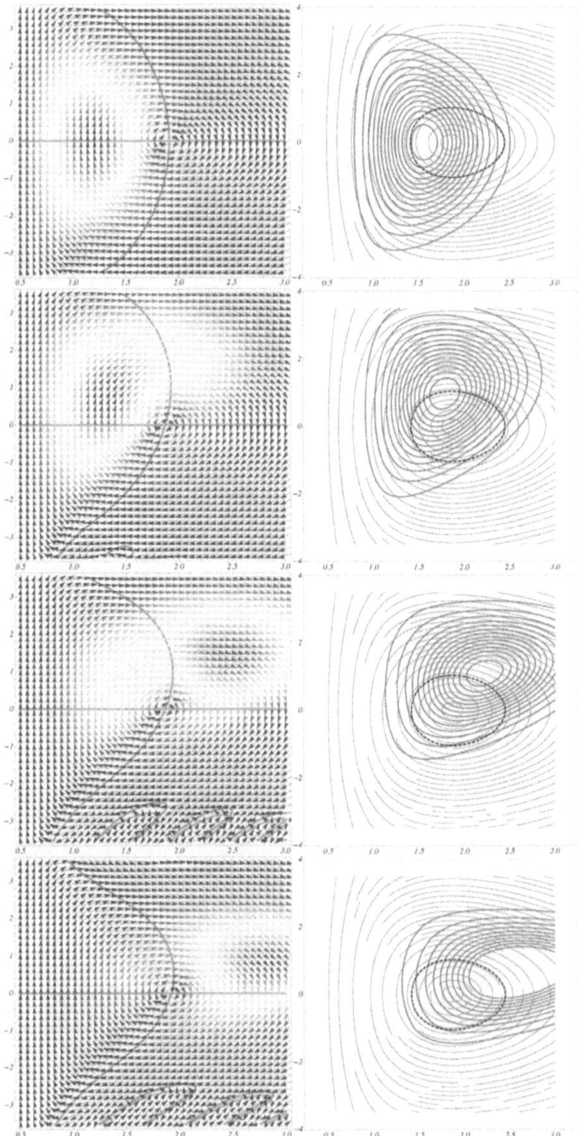

FIGURE 5.6: (Color online) First column: Time evolution of the normalized quantum Wigner flow fields, $\mathbf{J}/|\mathbf{J}|$, scaled by the *ThermometerColor* scheme (from zero (blue) to one (red)) in the $x - p$ plane. As before, the green (orange) contour lines are for $J^\alpha_{x(p)}(x, p; \tau) = 0$. Second column: Time evolution of the dominant region (red contours) of the Wigner function, $W^\alpha(x, p; \tau)$ superimposed by classical trajectories (thin black streamlines). The plots are for $\omega\tau = 0$, $\pi/4$, $\pi/2$, and $3\pi/4$, from top to bottom, with $E = 1/2$ and $\alpha = 7/2$.

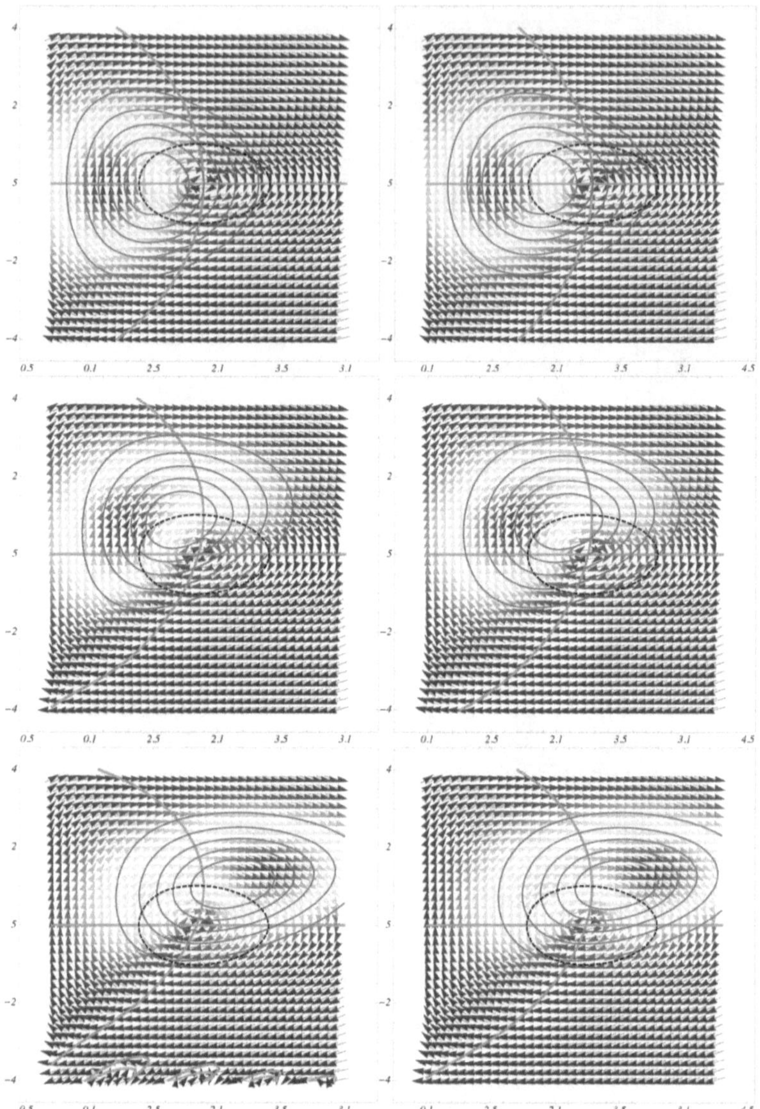

FIGURE 5.7: (Color online) Time evolution of the Wigner function, $W^\alpha(x, p; \tau)$, and the corresponding features of the Wigner flow in the phase-space ($x - p$ plane), for $\omega\tau = 0$ (first row), $\pi/4$ (second row), and $\pi/2$ (third row) (similar to Fig. 5.6), and $E = 1/2$ and $\alpha = 11/2$ (first column) and $\alpha = 15/2$ (second column). The quantum flows are in agreement with Fig. 5.6: the average values (and not the maximal probability values) for the canonical coordinates follow the classical trajectory. With respect to lower values of the parameter α, the quantum effects are suppressed.

by green and orange lines) emerge to brake the retarded evolution of the quantum flux. The classical profile does not exhibit such an overall locally compensation phenomena.

Finally, given that in the computation of the Wigner currents the quantum effects are accompanied by the non-linear contributions of the quantum potential, the exact result, Eq. (5.63), obtained from the infinite expansion from Eq. (5.58), guarantees that quantum corrections have been accurately accounted for in the above analysis.

The consistence of the above results can be assessed by comparison with analyses of the decoherence and classical correlations on the retrieval of classical behavior from the wave function of the Universe [179]. In principle, the investigation of quantum decoherence and of an emergence of classicality could be performed through a kind of wave packed spreading mechanism in the Schrödinger picture of quantum mechanics. Considering that this is rigorously suitable for the evolution of pure states, a broader description of how the loss of information takes place into the quantum to classical transition is more properly achieved through a density matrix description, for instance, through the Wigner-Weyl formalism, which provides a suitable picture of the quantum to classical transition. In this picture, decoherence takes place through a coarse-graining procedure [179, 180], which implies into the spreading of the Wigner functions (and associated coordinate probability densities), whereas coordinate and momentum classical correlations require a localized version of them [179]. That is, quantum coherence is lost through a coarse-graining procedure [166, 179] where either the density matrix is averaged over the phase-space variables or some external environment effect is considered. The decoherence triggers suppression of the correlation between coordinate and momentum, although they coexist as the quantum pattern is suppressed (but not destroyed) by sharply peaked quantum superpositions. That is the case of the HL cosmological system described along this section. The quasi-Gaussian sharpness of the resulting Wigner functions guarantees the correlation between averaged values of position and momenta, at the same time that the quantum pattern is still present in the adjacent quantum fringes, as it can be seen in Figs. 5.3 and 5.4. Hence, no decoherence effect takes place in this framework.

According to Hartle and Geroch prescriptions for quantum cosmology [181, 182], likewise in the quantum mechanics standard interpretation, peaks in the quantum (quasi-)distribution function are equivalent to predictions in quantum cosmology and, for the HL coherent peaked superposition discussed here, the resulting quantum dynamics is consistent with the classical correlation between position and momentum in a phase-space description.

Of course, it is worth mentioning that our description is achievable thanks to the mathematical manipulability of the Weyl transformed of the associated Laguerre polynomials that describe the HL eigensystem. Pure states corresponding to sharply peaked quasi-Gaussian wave functions are built without additional constraints, and these can be compared with those resulting from coarse-graining methods, required to set up the decoherence process. Given that no decoherence mechanism takes place in our framework, the quantum to classical transition is highly constrained by the choice of the HL cosmology, namely by the analytical manipulability provided by the WdW wave functions, and by the corresponding Wigner functions. In comparison, the WKB approximation and the coarse-graining procedure suggested in Ref. [179] is more general. In fact, a crucial point for the accuracy of the WKB analysis [179] is the significance of the $\mathcal{O}(\hbar)$ quantum corrections. This is irrelevant at our scenario of HL cosmology where all orders in $\hbar$ are absorbed in our procedure (cf. Eqs. (5.38), (5.43), and (5.62)). In the WKB approximation, quantum corrections are suppressed even though quantum interference fringes yield a Wigner function with a very large number of peaks which average to zero. In quantum cosmology, these cannot be neglected as their contribution yields observable averaged values. Therefore, in some cases, it suppresses the classical correlations. This is not the case at our HL approach, where the quasi-Gaussian quantum superposition in the phase-space guarantees the effectiveness of a WKB analysis[6]. However, the same cannot be asserted, for instance, to the following discussion of an analogous bounce model where, as one shall see, the presence of quantum interference effects is only completely captured by exact Wigner functions.

Wigner flow and quantum effects for bounce models

One of the fundamental questions in quantum cosmology concerns the initial singularity. Once the quantum cosmology in the minisuperspace framework admits a universe described by a wave function satisfying the WdW equation, some simple analytical extensions of the WdW solutions set constraints into the general features of the probability, time and ensued boundary conditions [111]. In fact, several hypothesis for circumventing the initial singularity have been suggested such as, for instance, the no-boundary and the tunneling proposals [87, 115–117].

For the HL models considered in this manuscript, an equivalent bounce model is obtained through the extension of the coordinate a (or x) from $(0, \infty)$ to $(-\infty, \infty)$, with a quasi-singularity at $a = 0$. It means that despite the presence of a potential barrier at $a = 0$, the wave functions from left to right are probabilistically connected. It can be implemented on the above obtained results by simply suppressing the step-functions $\Theta(x)$ from the integration Eqs (5.34)-(5.43).

---

[6]In particular, for the HL associated classical dynamics driven by the Hamiltonian Eq. (5.44), the classical time variable is identified with τ from the quantum framework, as to guarantee the reproduction of classical trajectories by the HL wave function quantum superpositions from Eqs. (5.28)-(5.30). Such a correspondence provides the elements for identifying τ with the WKB semiclassical time [123] in a generalized analysis where (cf. Ref. [?], pag. 402) the classical probability density is proportional to the time that the particle spends in an interval Δx such that the coarse-grained quantal probability density agrees with the classical probability density.

The corresponding result for the Wigner function should then read

$$
W_B^\alpha(x, p; \tau) \;=\; \frac{1}{\sqrt{\pi}} \frac{\Gamma(3/2 + \alpha)}{\Gamma(1 + \alpha)} (\mu x^2)^{1+\alpha} \exp\left(-\mu x^2\right)
$$

$$
\sum_{k=0}^{\frac{1}{2}+\alpha} \frac{x^{-(1+2k)}}{\Gamma(1+k)\,\Gamma(3/2+k+\alpha)} \frac{d^k}{d\mu^k} \left[\mu^{-1/2} \exp\left[-\frac{(p+\tilde\mu x)^2}{\mu} \right] \right] \tag{5.71}
$$

which, as can be seen from Fig. 5.8, it introduces some novel quantum features to its time evolution and the associated currents. In Fig. 5.8 the origin of quantum fluctuations on the right-hand side is due to the probabilistic connection to the left-hand side. Even with such quantum fluctuations, for semi-integer values of α one has exactly the same result for the purity of the associated quantum superpostion. As before, the local fluctuations do not affect the global evolution of the purity for such quantum superpositions.

5.5 Perturbative inclusion of the cosmological constant and the age of the Universe

Let one now turn back to the expressions for the Wigner currents, Eqs.(5.57)-(5.63). The inclusion of perturbative contributions due to ℓx^4 leads to the following additional contribution to the Wigner current,

$$
\begin{aligned}
J_p^{\alpha(\ell)}(x, p; \tau) &= \frac{\ell}{2}\left[2x^3 - \frac{2\times 3 \times 2}{2\times 3!} x \left(\frac{\partial}{\partial p}\right)^2 \right] W^\alpha(x, p; \tau) \\
&= \frac{\ell}{2}\left(2x^3\, W^\alpha(x, p; \tau) + \frac{2x}{\pi} \int_{-\infty}^{+\infty} dy\, y^2 \exp(2\,i\,p\,y)\, F_\alpha^*(x+y)\, F_\alpha(x-y) \right) \\
&= \frac{\ell}{2}\left(4x^3\, W^\alpha(x, p; \tau) - \frac{2x}{\pi} \int_{-\infty}^{+\infty} dy\, (x^2 - y^2)\, \exp(2\,i\,p\,y)\, F_\alpha^*(x+y)\, F_\alpha(x-y) \right) \\
&= 2\ell x^3 \left(W^\alpha(x, p; \tau) - \frac{1}{2x^2}\frac{1}{\pi} \int_{-\infty}^{+\infty} dy\, (x^2 - y^2)\, \exp(2\,i\,p\,y)\, F_\alpha^*(x+y)\, F_\alpha(x-y) \right) \\
&= 2\ell x^3 \left(W^\alpha(x, p; \tau) - \frac{1}{2\mu x^2}\frac{\alpha+2}{\Gamma(\alpha+1)} W^{\alpha+1}(x, p; \tau) \right).
\end{aligned} \tag{5.72}
$$

Since the background structure of $W^\alpha(x, p; \tau)$ does not change, the results due to Eq. (5.72) can only be considered perturbatively. Fig. 5.9 shows the time evolution of the Wigner flow (background arrows) for the perturbed HL quantum superposition in the presence of a cosmological constant parameterized by ℓ. Notice that lighter regions correspond to the probabilistically more relevant components of the Wigner flow bound by the unperturbed Wigner function, $W^\alpha(x, p; \tau)$, that moves periodically according to $\mu_{(\beta,\tau)}$ and $\tilde\mu_{(\beta,\tau)}$. The added perturbation due to the cosmological constant (proportional to minus x^4) yields conditions for quantum tunneling, which can be observed for times $\sim \omega\tau$ as depicted in Fig. 5.9 for $\ell = 0.022$. In the pictorial representation shown in Fig. 5.9, the Wigner function approaches the perturbative barrier (cf. Fig. 5.1) depicted by lighter regions which extend over a larger region of the phase-space. This means that part of its probabilistic contribution arises from

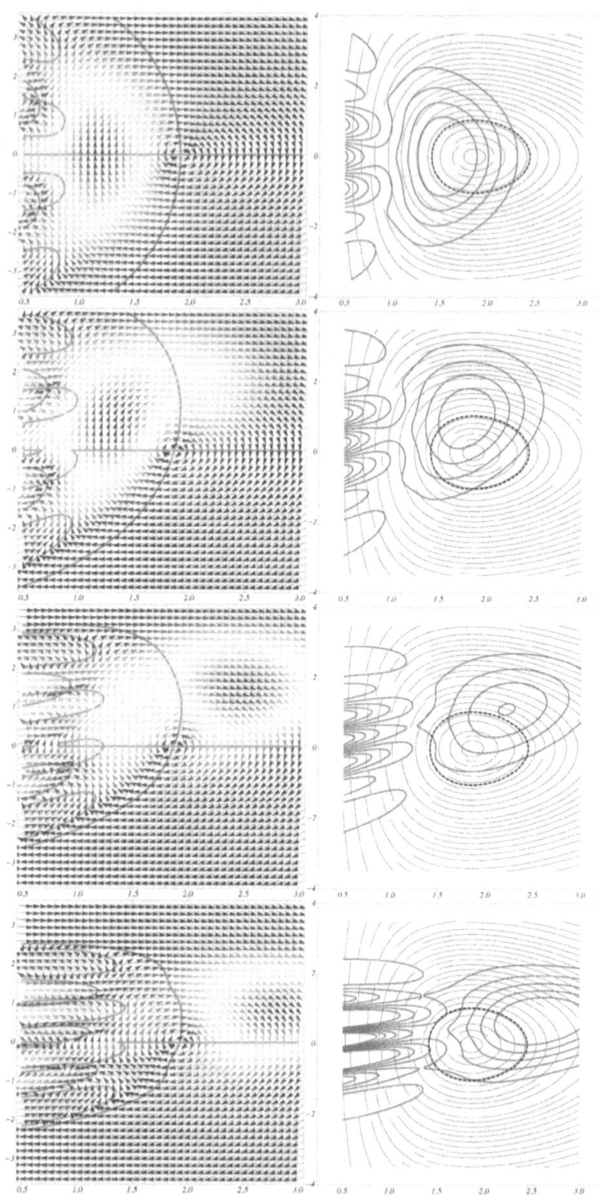

FIGURE 5.8: (Color online) Bounce HL cosmology with time evolution of the quasi-Gaussian Wigner function, $W^\alpha(x, p; \tau)$ in the phase-space $x - p$ plane. Plots are for $\omega\tau = 0$, $\pi/4$, $\pi/2$, and $3\pi/4$, from top to bottom (similar to Fig. 5.6), with $E = 1/2$ and $\alpha = 7/2$. The quantum features exhibited by the contour lines are in correspondence with Fig. 5.6.

tunneling (or transmission over the barrier) creating a novel cosmological phase driven by the perturbative current components, $J_p^{\alpha(\ell)}(x,\,p;\,\tau)$ (outside the barrier).

To quantitatively understand the above described dynamics [183], one can identify the total Wigner flow momentum direction component described by $J_p^{\alpha(Tot)}(x,\,p;\,\tau) = J_p^{\alpha}(x,\,p;\,\tau) + J_p^{\alpha(\ell)}(x,\,p;\,\tau)$, with the contributions for incident (background blue arrows for $p > 0$), transmitted (yellow arrows for $p > 0$) and reflected (orange arrows for $p < 0$) as depicted in Fig. 5.9, such that

$$
\begin{aligned}
\text{incident flow} \quad &\rightarrow \quad \Theta(+p)\,J_p^{\alpha}(x,\,p;\,\tau), \\
\text{transmitted flow} \quad &\rightarrow \quad \Theta(-p)\,J_p^{\alpha(\ell)}(x,\,p;\,\tau), \\
\text{reflected flow} \quad &\rightarrow \quad \Theta(-p)\,(J_p^{\alpha}(x,\,p;\,\tau) + J_p^{\alpha(\ell)}(x,\,p;\,\tau)).
\end{aligned}
$$

By observing that any component J_p^{α} has an even parity with respect to $p \rightarrow -p$, and that $\Theta(+p) + \Theta(-p) = 1$, with $\partial_p \Theta(+p) = -\partial_p \Theta(-p)$, one has

$$
\begin{aligned}
\int_{-\infty}^{+\infty} dp\,\partial_p J_p^{\alpha(Tot)} &= \int_{-\infty}^{+\infty} dp\,\partial_p[\Theta(+p)\,J_p^{\alpha} + \Theta(-p)\,(J_p^{\alpha} + J_p^{\alpha(\ell)}) + \Theta(+p)\,J_p^{\alpha(\ell)}] \\
&= \int_0^{+\infty} dp\,\partial_p J_p^{\alpha} - \int_0^{+\infty} dp\,\partial_p(J_p^{\alpha} + J_p^{\alpha(\ell)}) - \int_0^{+\infty} dp\,\partial_p J_p^{\alpha(\ell)}.
\end{aligned}
\tag{5.73}
$$

In addition, from the continuity equation for the momentum component (cf. Eq. (D.6) in the Appendix II), one has

$$
\frac{d}{d\tau}|G(p;\,\tau)|^2 = \int_0^{+\infty} dx\,\partial_p J_p(x,\,p;\,\tau),
\tag{5.74}
$$

which vanishes[7] after a symmetric integration over p. By substituting the result from Eq. (5.73) into the p-integrated version of Eq. (5.74) one has

$$
\int_0^{+\infty} dx \left[\int_0^{+\infty} dp\,\partial_p J_p^{\alpha} - \int_0^{+\infty} dp\,\partial_p(J_p^{\alpha} + J_p^{\alpha(\ell)}) - \int_0^{+\infty} dp\,\partial_p J_p^{\alpha(\ell)} \right] = 0,
\tag{5.75}
$$

where, in the integrand, the first term is associated to the incident probability which, of course, is shown to be equal to unity, the second term is associated to the reflection probability, R, and the last term is associated to the transmission probability, T, so that one consistently obtains $R + T = 1$.

---

[7]It has been fit to the adequate interval of $x \in (0,\infty)$.

The quantity T written in terms of

$$\begin{aligned}
T \equiv T(\omega\tau) &= \int_0^{+\infty} dx \int_0^{+\infty} dp\, \partial_p J_p^{\alpha(\ell)}(x, p; \tau) \\
&= \int_0^{+\infty} dx (J_p^{\alpha(\ell)}(x, \infty; \tau) - J_p^{\alpha(\ell)}(x, 0; \tau)) \\
&= -\int_0^{+\infty} dx\, J_p^{\alpha(\ell)}(x, 0; \tau)
\end{aligned} \tag{5.76}$$

defines the *transmission (or decaying) rate*, from which one can compute the age of the Universe.

By substituting Eq. (5.72) into Eq. (5.76) one obtains

$$\begin{aligned}
T(\omega\tau) &= \int_0^{+\infty} dx \int_0^{+\infty} dp\, \partial_p J_p^{\alpha(\ell)}(x, p; \tau) \\
&= -2\ell \int_0^{+\infty} dx\, x^3 \left(W^\alpha(x, 0; \tau) - \frac{1}{2\mu x^2} \frac{\alpha+2}{\Gamma(\alpha+1)} W^{\alpha+1}(x\,0; \tau) \right),
\end{aligned} \tag{5.77}$$

which results into a symbolic expression that be approximated by a constant value given by

$$T(\omega\tau) \simeq 2\ell \times 10^{\frac{3\alpha}{10}-2} = 2g_c^{1/2} \frac{g_\Lambda}{g_c^2} \times 10^{\frac{3\alpha}{10}-2} = \frac{g_c^{1/2}}{18\pi^2} \frac{\Lambda}{M_{\rm Pl}^2} \times 10^{\frac{3\alpha}{10}-2}, \tag{5.78}$$

where, in the last step, the quantities have been written in terms of Planck units and

$$\frac{g_\Lambda}{g_c^2} = \frac{1}{36\pi^2} \frac{\Lambda}{M_{\rm Pl}^2}.$$

Given that the transmission probability through the barrier is provided by the cosmological constant contribution to the cosmic inventory, Ω_Λ, then

$$T(\omega\tau) \simeq \frac{\Omega_\Lambda}{\omega\tau} = \frac{1}{2g_c^{1/2}} \frac{\Omega_\Lambda}{\tau}. \tag{5.79}$$

The contribution, Ω_Λ, is estimated to be

$$\Omega_\Lambda = \frac{g_c\, g_0}{36\pi^2} \times 10^{\frac{3\alpha}{10}-2} \times \tau_{Age}, \tag{5.80}$$

where $\Lambda = g_0 M_{\rm Pl}^2/2$. Finally, as to recover the current (phenomenological) age of the Universe, $\tau_{Age} \simeq 8 \times 10^{60} T_{\rm Pl}$, for $\Omega_\Lambda \simeq 0.7$, from typical values, $g_c \simeq 1$ and $g_0 \simeq 10^{-123}$, one should have $10^{\frac{3\alpha}{10}-2} \simeq 1.5 \times 10^{61}$, which leads to $\alpha \sim 210$ and is not affected the contribution from *radiation* and *stiff matter* components since $\Omega_R \sim \sqrt{\Omega_S}$ for large values of α.

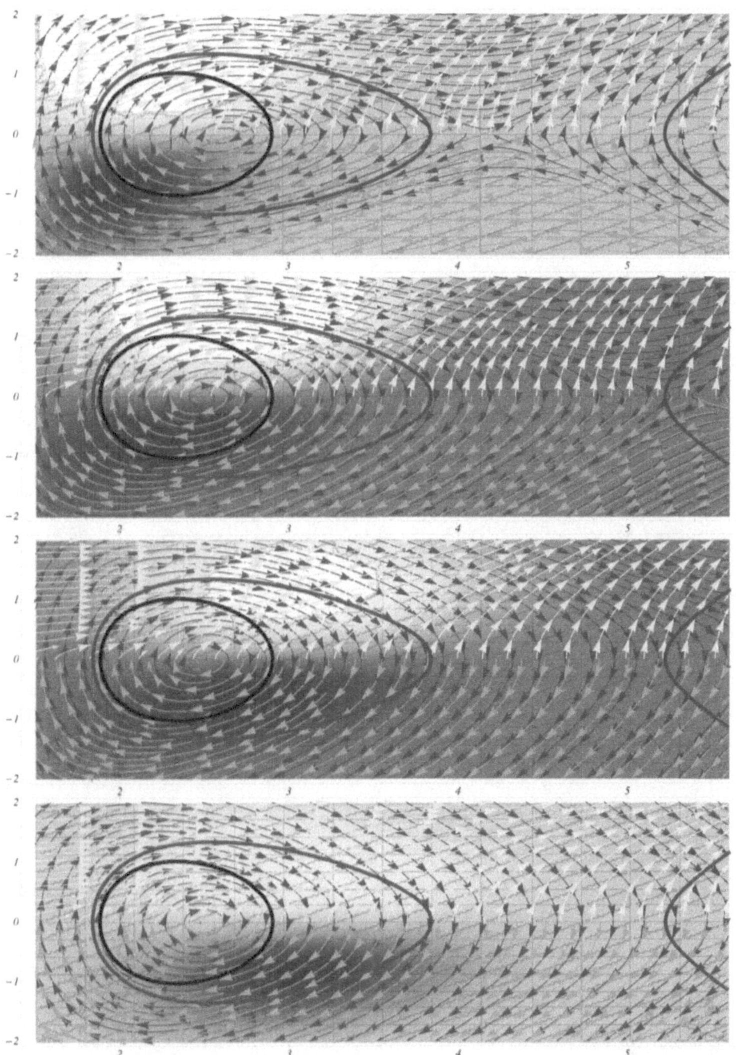

FIGURE 5.9: (Color online) Complete profile of the Wigner flow (streamlines) for the perturbed HL quantum system in the presence of a cosmological constant. The Wigner function background color scheme denotes higher values for lighter regions and lower values darker regions. The amplitude of the Wigner function modulates the (unscaled) arrows. Streamlines indicate the Wigner flow (normalized to each point) for *transmitted* (yellow arrows, $p > 0$) and *reflected* (orange arrows, $p < 0$) currents (since they are normalized to each point, the streamline plot corresponds to values of $\mathbf{J}/|\mathbf{J}|$, therefore only the flow direction is relevant). Results are for $E = 2$, $\alpha = 11/2$, and $\ell = 0.022$ (for convenience) and for $\omega\tau = \pi/4$, $\pi/2, 3\pi/4$ and π, from top to bottom.

Chapter 6

Conclusions

In this book various issues concerning PSNCQM were studied. The fundamental tenets related to the reproducibility of quantum states through quantum cloning and quantum teleportation were examined in the framework of phase-space NC QM in Chapter 2. The no-cloning theorem was cast into the Wigner formalism for QM and then generalized for phase-space NC QM, showing that a perfect copy of a state is impossible to be obtained within the NC framework as well. It is seen that the only feature required for this result to hold is the unitarity, which is also a feature of NC QM.

In what concerns quantifiers of the above mentioned quantum processes, the quantum fidelity for continuous variable teleportation protocols in the NC phase-space was computed using the NC Wigner function definitions (cf. Ref. [15]). It has been shown that, when computed for 2-dim Gaussian and HO *input* states, the noncommutativity does not affect the fidelity results of duplicated states. In fact, the results for quantum fidelity are shown to be independent of the SW map which relates NC and ordinary quantum systems. This is in line with results of Ref. [15], where it is shown that physical results such as expectation values and transition amplitudes are independent of the SW map.

Finally, a procedure for quantum teleportation of continuous variables in the phase-space NC QM was developed as to take into account some subtle modifications on the commutative teleportation protocol. These modifications amounted to change the preparation of the entangled state in order to account for the additional NC relations so to agree with the uncertainty principles imposed by the deformed HW algebra. To summarize, NC QM does impose, likewise in QM, relevant obstacles to the reproduction of states as well as associated duplication protocols.

In Chapter 3 the deformation of the relativistic dispersion relation arising from the breaking of Lorentz symmetry in phase-space NCQM was analysed. For this purpose, the framework of phase-space as a symplectic manifold was used, as well as its extension to include time as a coordinate and not as a parameter. This leads to the extended phase-space formalism. The deformation of the HW algebra of observables was used to compute the deformation in the symplectic form defined in the extended phase-space. This gives rise to fundamental length and momentum scales, that results in the breaking of Lorentz invariance. This fact, together with the symplectic point of view for the phase-space allows for the use of Darboux's map to find the deformation of the dispersion relation for a relativistic particle. The new relation was tested with GRB data from several sources at different redshifts and the obtained constraint for the momentum scale is $\sqrt{\eta} \lesssim 10^{-12}\,\mathrm{eV/c}$. This is a very stringent bound and improves on previous low-energy physics results, $\sqrt{\eta} \lesssim 10^{-5}\,\mathrm{eV/c}$ [24, 25].

Additionally, a pseudo-Riemannian metric was introduced in the extended noncommutative phase-space, via a compatible triplet comprised of a symplectic form, an almost complex structure and a metric tensor. This allows for a straightforward notion of distance in phase-space that arises naturally from compatibility requirements on these three structures. With this framework, the noncommutative metric tensor is constructed. It is found that the naive commutative limit $\theta, \eta \to 0$ cannot be taken since it renders the metric ill defined. Rather, this limit must be considered for fixed θ/η ratio. With this, one recovers the commutative limit iff $\theta = \eta$. The significance of this result, as well as its implications clearly deserve further research and it is expected that they may appear in high energy and statistical physics.

In Chapter 4, a spherically symmetric null-like hypersurface collapsing into a BH is considered. The classical and quantum framework for the problem are constructed through the assumptions that inside and outside regions of the shell are mapped into two distinct KS metrics. The junction condition formalism for null hypersurfaces is applied in order to obtain the action for the system. Furthermore, since this hypersurface is a null boundary for both KS spacetimes, an extra boundary term for the action is admitted [106, 107]. This allows for the definition of the Hamiltonian of the system and for obtaining the WdW equation for the quantum mechanical problem. A more general approach for the description of collapsing shells dynamics can be found in Ref. [184], in particular, in what concerns rotational degrees of freedom. We point out that our approach leads to the same results for the discussed cases.

The WdW equation is solved, and the wave function is found to be given by a family of modified Bessel functions, which exhibit an oscillatory behavior in the limit corresponding to the BH singularity, $\Omega \to \infty$. The squared modulus of this wave function does not depend on β, thus, from the map, Eq. (4.4), there is a symmetry with respect to the raduis $t = M$. Therefore, the limit $\Omega \to \infty$ corresponds to both the singularity and the event horizon.

Extending the Heisenberg-Weyl algebra and considering the phase space noncommutativity, it is shown that the configuration space θ parameter contributes only as a shift in the wave function of the commutative problem. On the other hand, when the NC parameter η is considered, a damped wave function arises for $\Omega \to \infty$. It is found that the squared modulus of the wave function does not depend on β, thus the probability density vanishes as $t \to 0$ and $t \to 2M$. Thus, the introduction of the momentum noncommutativity acts as a regulator for the wave function, a feature previously encountered for the BH problem without the shell [27, 94]. This result suggests that a more complex NC framework is required to obtain square integrable wave functions that vanish at the singularity as found in Ref. [26].

Chapter 5 is devoted to the study of the Wigner function and the corresponding Wigner flow for the HL quantum cosmology in the minisuperspace approach, so to visualize the transition from quantum to classical cosmological behaviors in the presence of radiation, curvature and stiff matter components. In particular, it has been shown that a quantum mechanical parameterization of time as the variable canonically conjugated to the radiation energy density contribution to the HL Hamiltonian is consistent with the classical time evolution.

In general, the classical limit for quantum cosmologies described by HL minisuperspace models are difficult to obtain due to the pattern of oscillations of the Wigner functions. To study the transition from quantum to classical cosmology, the HL model has been described in terms of its Wigner currents. Our result shows that the averaged quantities obtained from the quasi-Gaussian Wigner function built from the solutions of the WdW equation for the HL quantum mechanical problem coincide with the equivalent classical trajectory of the Universe. In particular, the classical trajectory matches almost perfectly the maxima of the peak of the Wigner function, with increasingly high accuracy for higher values of the parameter α, associated to the stiff matter contribution.

Besides providing an identification with the corresponding classical cosmology, the exact expressions for the Wigner currents show an approximated Liouvillian character for the Wigner flow. In addition, a cosmological constant contribution breaks the oscillatory behavior of the Wigner flow due to radiation, curvature and stiff matter contributions and introduces a novel quantum decaying scenario which allows for estimating the age of the Universe. Moreover, an extension for a kind of bounce model, which expands the space-like coordinate limit from $-\infty$ to $+\infty$, reveals an interesting pattern of quantum interference that produces an identifiable distortion of the Wigner flow.

Finally, it is worth mentioning that the procedure discussed here can be extended to other quantum cosmological scenarios as, for instance, to Kantowski-Sachs [92, 93] and modular quantum cosmologies [185]. Furthermore, the problem of reaching the classical limit through coarse-graining arguments can be considered by the inclusion of additional friction and diffusion terms in extended versions of our proposal through a deformed quantization formalism.

Appendix A

Appendix A

A.1 Schwarzschild solution

The Schwarzschild metric is a solution to the Einstein field equations (EFE) with spherical symmetry and vanishing energy-momentum tensor everywhere, i.e., in vacuum. In units where $G = 1$ and $c = 1$, it is given by:

$$ds^2 = -\left(1 - \frac{2M}{r}\right) dt^2 + \left(1 - \frac{2M}{r}\right)^{-1} dr^2 + r^2 \left(d\theta^2 + \sin^2\theta \, d\varphi^2\right). \tag{A.1}$$

where M is the mass contained inside the event horizon, that is, the mass of the BH: in the case of a collapsed star or shell it is the mass of the physical object in question. It is easy to see that this metric is singular at $r = 2M$ and $r = 0$. It is known that the singularity at $r = 2M$ is not essential and can be removed by a suitable change of coordinates. A particularly useful set of coordinates is called Eddington-Finkelstein (EF) coordinates. These are coordinates defined for observers following radial null trajectories, hence:

$$ds^2 = 0 \Leftrightarrow dt^2 = \left(1 - \frac{2M}{r}\right)^{-2} dr^2 := (dr^*)^2, \tag{A.2}$$

which thus defines the null radial coordinate as:

$$r^* = r + 2M\ln\left|\frac{r - 2M}{2M}\right|, \tag{A.3}$$

which is known as the Regge-Wheeler radial coordinate. With this coordinate in mind we notice that we can define two coordinates for which the invariant interval vanishes:

$$v := t + r^*, \qquad u := t - r^*, \tag{A.4}$$

each of which ranges from $-\infty$ to $+\infty$. We can then rewrite the Schwarzschild metric in ingoing Eddington-Finkelstein coordinates (v, r, θ, ϕ) as:

$$ds^2 = -\left(1 - \frac{2M}{r}\right) dv^2 + 2dvdr + r^2d\Omega^2 \tag{A.5}$$

wherein $r = r(v)$. It is obvious that no singularity exists at r=2M because of the non-diagonal factor of the metric. Although this coordinates are initially only defined for $r > 2M$, since there is no singularity at $r = 2M$ we can extend them for $r > 0$. However singularity at the origin still remains, since it is an essential singularity, which can be verified through the components of the Riemann tensor, which are infinite at this point. This set of coordinates describes a part of spacetime called a black hole. This is easily seen by computing the light cone for a radially falling particle: for $r < 2M$ the light cone does not allow for the coordinate r to increase, thus prohibiting any falling object from escaping this region. Alternatively we can write tis metric in outgoing Eddington-Finkelstein coordinates (u, r, θ, ϕ):

$$ds^2 = -\left(1 - \frac{2M}{r}\right) du^2 - 2dudr + r(u)^2d\Omega^2. \tag{A.6}$$

This is similar to the ingoing coordinates, but this time, for $r < 2M$ the light cone forces particles to follow trajectories of increasing r, hence covering a part of spacetime designated as *White Hole* (WH). It must be noted that the change of coordinates and their extension past the limit in Schwarzschild coordinates allows for the mapping of different spacetime regions of the same solution to the EFE. Thus, selection of coordinates is of prime importance in GR. Additionally to the EF coordinates, there is another set of coordinates, named Kruskal-Szekeres (KS) coordinates, which are defined by:

$$U := -e^{-u/4M}, \qquad V := e^{v/4M}, \tag{A.7}$$

where u and v are EF coordinates. Since these are defined for $r > 2M$, the new U, V coordinates are defined for $U < 0$ and $V > 0$, respectively. With these we can now rewrite the metric Eq. (A.1) as:

$$ds^2 = -\frac{32M^3}{r}e^{-r/2M}dUdV + r^2d\Omega, \tag{A.8}$$

where $r = r(U, V)$ is defined by:

$$UV = -\frac{r - 2M}{2M}e^{r/2M}. \tag{A.9}$$

Since there is no singularity at $r = 2M$, i.e., $V = 0$ or $U = 0$, in these coordinates, one can extend them to $U > 0$ and $V < 0$. Doing this enables the description of the full manifold with one coordinate system: for $V > 0$ and $U < 0$ the outside of the event horizon is described;

the inside of the BH is described by $U > 0$ and $V > 0$; the WH appears at $U < 0$ and $V < 0$; and finally, for $U > 0$ and $V < 0$ the is a spacetime region, similar to the outside of the event horizon, but in which time flows in the opposite direction. The manifold equipped with the KS coordinates are called the Kruskal manifold and is the maximal analytic extension of the Schwarzschild solution.

Appendix B

Appendix B

In Sec. 4.2 it was argued that one could take the coordinate, t, to be the same on both parts of the manifold. However, this is not necessarily true and the same result, namely the vanishing of the boundary terms, can be recovered by assuming that the time coordinates are different. Indeed, starting from metrics of the form:

$$ds_\pm^2 = e^{\sqrt{3}\beta_\pm}\, du \left(e^{\sqrt{3}\beta_\pm}\, du + 2\zeta N \pm dt_\pm \right) + e^{-2\sqrt{3}\Lambda} \left(d\theta^2 + \sin^2\theta\, d\phi^2 \right), \tag{B.1}$$

one can perform a change of coordinates, $T_\pm := e^{-2\sqrt{3}\Lambda_\pm}$, so to obtain:

$$ds_\pm^2 = e^{\sqrt{3}\beta_\pm}\, du \left(e^{\sqrt{3}\beta_\pm}\, du - \frac{2\zeta N}{\sqrt{3}T_\pm\,\alpha_\pm}\, dT_\pm \right) + T_\pm^2 \left(d\theta^2 + \sin^2\theta\, d\phi^2 \right), \tag{B.2}$$

where $\alpha_\pm(T) = \dot\Lambda_\pm(T)$. The holonomic vectors are then given by:

$$e^\mu_{(A)} := \frac{\partial x^\mu}{\partial y^A} = \delta^\mu_{A'} \tag{B.3}$$

for $A = \theta, \phi$, and by

$$e^\mu_{(T)} := \frac{\partial x^\mu}{\partial T} = \partial_T, \tag{B.4}$$

for $T_\pm$. These definitions allow for the computation of the induced metric using $g_{ab} = g_{\mu\nu}e^\mu_{(a)}e^\nu_{(b)}$ for both sides as:

$$ds_{|\Sigma}^2 = T_\pm^2\, d\Omega^2. \tag{B.5}$$

For the hypersurface to possess a well defined structure, the induced metrics on each of its sides must coincide, and the coordinates $T_\pm$ must be the same. In the coordinate system considered in the body of the paper (with t rather than T), this leads to the condition:

$$\Lambda_+(t_+) = \Lambda_-(t_-). \tag{B.6}$$

Besides providing the condition for the junction of the spacetimes, this coordinate system also allows for a convenient set of coordinates to be introduced on the shell, the coordinates (T, θ, ϕ). These are the natural coordinates on Σ and they are the same on both sides of the hypersurface. Using the calculations performed in Sec. 4.2 alone, in the novel coordinate system, Eq. (B.4), the normal and transversal vectors to Σ are then given by:

$$n_\pm^\mu = -\frac{\zeta\, T\, \alpha_\pm e^{-\sqrt{3}\beta_\pm}}{N_\pm \chi_\pm}\, \partial_T,$$
(B.7)

$$M_\pm^\mu = \frac{\chi_\pm}{\varepsilon_\pm}\partial_u - \frac{\zeta\, T\, \alpha_\pm e^{\sqrt{3}\beta_\pm}\chi_\pm}{2\varepsilon_\pm N_\pm}\,\partial_T.$$
(B.8)

The projections of these vectores onto the three dimensional hypersurface are:

$$n_\pm^a = -\frac{\zeta\, T\, \alpha_\pm e^{-\sqrt{3}\beta_\pm}}{N_\pm \chi_\pm}\, \partial_T,$$
(B.9)

for the normal vector, and

$$M_a^\pm = -\frac{\zeta N_\pm e^{\sqrt{3}\beta_\pm}\chi_\pm}{\varepsilon_\pm\, T\, \alpha_\pm}\, \partial_T,$$
(B.10)

for the transverse one. The requirement that the quantities n^a and M_a coincide on both sides of the shell determines a unique structure for Σ. This condition imposes restrictions on the normalization factors, namely:

$$\left.\frac{e^{-\sqrt{3}\beta_-}\alpha_-}{N_-\chi_-}\right|_{u=u_0} = \left.\frac{e^{-\sqrt{3}\beta_+}\alpha_+}{N_+\chi_+}\right|_{u=u_0},$$
(B.11)

and,

$$\left.\frac{N_- e^{\sqrt{3}\beta_-}\chi_-}{\varepsilon_-\alpha_-}\right|_{u=u_0} = \left.\frac{N_+ e^{\sqrt{3}\beta_+}\chi_+}{\varepsilon_+\alpha_+}\right|_{u=u_0}.$$
(B.12)

Once again, these lead to the conclusion that $\varepsilon_- = \varepsilon_+$. The next step is to compute the transverse curvature, which is found to be,

$$K_{tt}^\pm = \frac{\zeta e^{\sqrt{3}\beta_\pm}\chi_\pm}{\varepsilon\, T^2\, \alpha_\pm^2}\left[T\alpha_\pm \dot{N}_\pm + N_\pm\left(\alpha_\pm\left(\sqrt{3}T\dot{\beta}_\pm - 1\right) - T\dot{\alpha}_\pm\right)\right],$$

$$K_{\theta\theta}^\pm = \frac{\zeta\, T^2\, \alpha_\pm e^{\sqrt{3}\beta_\pm}\chi_\pm}{2\varepsilon N_\pm},$$
(B.13)

$$K_{\phi\phi}^\pm = K_{\theta\theta}^\pm \sin^2(\theta),$$

as well as the discontinuity in the curvature scalar,

$$
\begin{aligned}
R_{|\Sigma} = & -\frac{2\zeta e^{-\sqrt{3}\beta_-}}{\varepsilon N_-^2 \chi_-}\left[N_-\left(\alpha_-\left(\sqrt{3}T\dot{\beta}_- - 1\right) - T\dot{\alpha}_-\right) + \alpha_- T\dot{N}_-\right] + \\
& +\frac{2\zeta e^{-\sqrt{3}\beta_+}}{\varepsilon N_+^2 \chi_+}\left[N_+\left(\alpha_+\left(\sqrt{3}T\dot{\beta}_+ - 1\right) - T\dot{\alpha}_+\right) + \alpha_+ T\dot{N}_+\right],
\end{aligned}
\tag{B.14}
$$

where Eq. (B.11) was used in order to organize the expression. Aiming to compute the boundary action, we are left with the task of computing the trace of the Θ tensor and the non affinity parameter κ. In this setup their respective expressions are the following:

$$
\Theta_\pm = -\frac{2\zeta \alpha_\pm e^{-\sqrt{3}\beta_\pm}}{N_\pm \chi_\pm}, \qquad \kappa_\pm = \frac{\zeta T \alpha_\pm e^{-\sqrt{3}\beta_\pm}\dot{\chi}_\pm}{N_\pm \chi_\pm^2}.
\tag{B.15}
$$

Finally, once these results are inserted into the piece of the Lagrangian corresponding to the boundary, Σ, it vanishes and the boundary terms in the action are suppressed. Thus, the result of Sec. 4.2 remains unchanged even under more general assumptions.

Appendix C

Appendix C

From Eq. (5.34), the normalization condition over $W^\alpha(x, p; \tau)$ can be verified through some simple mathematical manipulations. Firstly, one notices that

$$\int_0^\infty dx \int_{-\infty}^{+\infty} dp\, W^\alpha(x, p; \tau) = \frac{2\,\mu^{1+\alpha}}{\pi\,\Gamma(1+\alpha)} \int_0^\infty dx\, x^{2(1+\alpha)} \exp\left(-\mu\, x^2\right) \int_{-\infty}^{+\infty} dp \exp\left(2\,i\,x\,p\,s\right) \times$$
$$\int_{-1}^{+1} ds\,(1-s^2)^{\frac{1}{2}+\alpha} \exp\left(-\mu\, x^2\, s^2\right) \exp\left(2i\,\tilde{\mu}\, x^2\, s\right), \quad \text{(C.1)}$$

where y has been parameterized as $y = x\,s$ and μ and $\tilde{\mu}$ follows from Eqs. (5.31) and (5.35), respectively. By observing that

$$\int_{-\infty}^{+\infty} dp \exp\left(2\,i\,x\,p\,s\right) = 2\pi\,\delta(2\,x\,s) = \frac{\pi}{|x|}\delta(s), \quad \text{(C.2)}$$

after substitution into Eq. (C.1), an integration over the variable s yields

$$\frac{\pi}{|x|} \int_{-1}^{+1} ds\,\delta(s)(1-s^2)^{\frac{1}{2}+\alpha} \exp\left(-\mu\, x^2\, s^2\right) \exp\left(2i\,\tilde{\mu}\, x^2\, s\right) = \frac{\pi}{|x|}, \quad \text{(C.3)}$$

and then

$$\int_0^\infty dx \int_{-\infty}^{+\infty} dp\, W^\alpha(x, p; \tau) = \frac{2\,\mu^{1+\alpha}}{\Gamma(1+\alpha)} \int_0^\infty dx\, x^{(1+2\alpha)} \exp\left(-\mu\, x^2\right) = 1. \quad \text{(C.4)}$$

By following a similar strategy, given the purity (cf. Eq. (5.27)),

$$\mathcal{P} = 2\pi \int_0^\infty dx \int_{-\infty}^{+\infty} dp\, \left(W^\alpha(x, p; \tau)\right)^2, \quad \text{(C.5)}$$

one notices that

$$\int_0^\infty dx \int_{-\infty}^{+\infty} dp \; (W^\alpha(x, p; \tau))^2 = \frac{4\,\mu^{2(1+\alpha)}}{\pi^2\,\Gamma^2(1+\alpha)} \times$$
$$\int_0^\infty dx\, x^{4(1+\alpha)} \exp\left(-2\,\mu\,x^2\right) \int_{-\infty}^{+\infty} dp \, \exp\left(2\,i\,x\,p\,(s+r)\right) \times$$
$$\int_{-1}^{+1} ds \int_{-1}^{+1} dr\, [(1-r^2)(1-s^2)]^{\frac{1}{2}+\alpha} \exp\left(-\mu\,x^2\,(r^2+s^2)\right) \exp\left(2i\,\tilde\mu\,x^2\,(s+r)\right) \quad \text{(C.6)}$$

After substituting

$$\int_{-\infty}^{+\infty} dp \, \exp\left(2\,i\,x\,p\,(s+r)\right) \;=\; 2\pi\,\delta(2\,x\,(s+r)) = \frac{\pi}{|x|}\delta(s+r) \qquad \text{(C.7)}$$

into Eq. (C.6), an integration over the variable r gives

$$\int_0^\infty dx \int_{-\infty}^{+\infty} dp \; (W^\alpha(x, p; \tau))^2 = \frac{4\,\mu^{2(1+\alpha)}}{\pi^2\,\Gamma^2(1+\alpha)} \times$$
$$\int_0^\infty dx\, x^{(3+4\alpha)} \exp\left(-2\,\mu\,x^2\right) \int_{-1}^{+1} ds\,(1-s^2)^{1+2\alpha} \exp\left(-2\,\mu\,x^2\,s^2\right). \quad \text{(C.8)}$$

By evaluating the integral over x, one has

$$\int_0^\infty dx\, x^{(3+4\alpha)} \exp\left(-2\,\mu\,x^2\,(1+s^2)\right) = \frac{1}{2^{3+2\alpha}\mu^{2(1+\alpha)}} \frac{\Gamma(2(1+\alpha))}{(1+s^2)^{2(1+\alpha)}}, \qquad \text{(C.9)}$$

which can be substituted into Eq (C.8) as to give

$$\begin{aligned}
\int_0^\infty dx \int_{-\infty}^{+\infty} dp \; (W^\alpha(x, p; \tau))^2 &= \frac{1}{2^{1+2\alpha}\,\pi} \frac{\Gamma(2(1+\alpha))}{\Gamma^2(1+\alpha)} \int_{-1}^{+1} ds\, \frac{(1-s^2)^{1+2\alpha}}{(1+s^2)^{2(1+\alpha)}} \\
&= \frac{1}{2^{2+2\alpha}\,\pi} \frac{\Gamma(2(1+\alpha))}{\Gamma^2(1+\alpha)} \frac{\sqrt{\pi}\,\Gamma(1+\alpha)}{2\Gamma(3/2+\alpha)} \\
&= \frac{1}{2\pi},
\end{aligned} \qquad \text{(C.10)}$$

that confirms that $\mathcal{P} = 1$ for $W^\alpha(x, p; \tau)$ defined by Eq. (5.34).

Appendix D

Appendix D

Interesting quantum aspects of a physical system can be revealed by the time evolution of the Wigner function [173, 186], $W(\xi_x, \xi_p; t)$, when it is cast in the form of a vector flux $\mathbf{J}(\xi_x, \xi_p; t)$ [187–189]. This flow drives the quasi-probability density in the phase-space as well as it reproduces the dynamics of a quantum system. Written in the form of a continuity equation [170, 173, 180]

$$\frac{\partial W}{\partial t} + \boldsymbol{\nabla}_\xi \cdot \mathbf{J} = \frac{\partial W}{\partial t} + \frac{\partial J_x}{\partial \xi_x} + \frac{\partial J_p}{\partial \xi_p} = 0, \tag{D.1}$$

through the $\xi_x - \xi_p$ decomposition, $\mathbf{J} = J_x\,\hat{\xi}_x + J_p\,\hat{\xi}_p$, with

$$J_x(\xi_x, \xi_p; t) \;=\; \frac{\xi_p}{m}\,W(\xi_x, \xi_p; t), \tag{D.2}$$

$$J_p(\xi_x, \xi_p; t) \;=\; -\sum_{k=0}^{\infty}\left(\frac{i\hbar}{2}\right)^{2k}\frac{1}{(2k+1)!}\left(\frac{\partial}{\partial \xi_x}\right)^{2k+1}V(\xi_x)\left(\frac{\partial}{\partial \xi_p}\right)^{2k}W(\xi_x, \xi_p; t), \tag{D.3}$$

the probability density, $|F(\xi_x; t)|^2$, and the momentum distribution $|G(\xi_p; t)|^2$, are related one to each other by

$$G(\xi_p; t) = (2\pi\hbar)^{-1/2}\int_{-\infty}^{+\infty}d\xi_x\,F(\xi_x; t)\,\exp(i\,\xi_p\,\xi_x/\hbar), \tag{D.4}$$

and have a time evolution given by

$$\frac{d}{dt}|F(\xi_x; t)|^2 = \int_{-\infty}^{+\infty}d\xi_p\,\partial_{\xi_x}J_x(\xi_x, \xi_p; t) = \partial_{\xi_x}j_x(\xi_x; t), \tag{D.5}$$

$$\frac{d}{dt}|G(\xi_p; t)|^2 = \int_{-\infty}^{+\infty}d\xi_x\,\partial_{\xi_p}J_p(\xi_x, \xi_p; t) = \partial_{\xi_p}j_p(\xi_p; t). \tag{D.6}$$

Of course, the first of the above equations has a quantum analog in the coordinate representation, that is

$$\int_{-\infty}^{+\infty}dp\,J_x(\xi_x, \xi_p; t) = j_x(\xi_x; t).$$

For the purpose of comparing classical and quantum dynamics, one firstly identifies the classical Hamiltonian phase-space velocity, $\dot{\boldsymbol{\zeta}} = \mathbf{v}_\zeta = (v_x, v_p)$, associated to a coordinate vector $\boldsymbol{\zeta} = (\xi_x, \xi_p)$, so to have the classical flow field given by $\mathbf{J} = \mathbf{v}_\zeta W$, with $v_x = \dot{\xi}_x = \xi_p/m$ and $v_p = \dot{\xi}_p = -\partial V/\partial \xi_x$. The analogy with fluid dynamics is indeed much more intuitive in the classical regime for which Eq. (D.3) reduces to the Liouville equation. In particular, the classical velocity for time independent conservative systems is divergence free, i.e. $\boldsymbol{\nabla}_\zeta \cdot \mathbf{v}_\zeta = 0$. In this case, to compute time variation of the integrated probability for a volume of fluid bound by an enclosing path, $\mathcal{C}$, which moves with $\mathbf{v}_{\zeta(\mathcal{C})} = (\xi_p/m, -\partial V/\partial \xi_x)$, Eq. (D.1) is cast in the form of

$$\frac{\partial W}{\partial t} + \boldsymbol{\nabla}_\zeta \cdot (\mathbf{v}_{\zeta(\mathcal{C})}\, W) \;=\; 0, \qquad (classical). \tag{D.7}$$

By observing that the convective derivative operator [173] corresponds to

$$\frac{D}{Dt} \equiv \frac{\partial}{\partial t} + \mathbf{v}_\zeta \cdot \boldsymbol{\nabla}_\zeta, \tag{D.8}$$

one can rewrite the continuity equation, Eq. (D.7), as

$$\frac{DW}{Dt} = -W\, \boldsymbol{\nabla}_\zeta \cdot \mathbf{v}_\zeta, \tag{D.9}$$

which corresponds to a conservation law whenever $\frac{DW}{Dt} = 0$. It establishes that the fluid-analog is Liouvillian and incompressible.

Otherwise, for the quantum case described by Eq. (D.1), $\mathbf{J}$ may be associated to $\mathbf{u}\,W$, and a typical non-Liouvillian [190] flow can be recognized through the divergence pattern of $\mathbf{u}$, $\boldsymbol{\nabla}_\zeta \cdot \mathbf{u} \neq 0$. The Wigner phase-velocity, $\mathbf{u}$, the quantum analog of $\mathbf{v}_{\zeta(\mathcal{C})}$, exhibits a subtle unbound divergent behavior expressed by

$$\boldsymbol{\nabla}_\zeta \cdot \mathbf{u} = \frac{W\, \boldsymbol{\nabla}_\zeta \cdot \mathbf{J} - \mathbf{J} \cdot \boldsymbol{\nabla}_\zeta W}{W^2}, \tag{D.10}$$

where $\boldsymbol{\nabla}_\zeta \cdot \mathbf{J} = W\boldsymbol{\nabla}_\zeta \cdot \mathbf{u} + \mathbf{u} \cdot \boldsymbol{\nabla}_\zeta W$. The condition that sets $\boldsymbol{\nabla}_\zeta \cdot \mathbf{u} \neq 0$ is quite helpful in identifying an approximated Liouvillian dynamics in the phase-space.

Besides producing a picture of the local quantum distortions, the operator $\boldsymbol{\nabla}_\zeta \cdot \mathbf{u}$ has also a closed relation with the rate of change of the purity [191] as given by [178]

$$\frac{1}{2\pi}\frac{D\mathcal{P}}{Dt} + \langle W\, \boldsymbol{\nabla}_\zeta \cdot \mathbf{u} \rangle = 0. \tag{D.11}$$

The rate of change of $\mathcal{P}$ is driven by quantum distortions over the background Liouvillian flow. From Eq. (D.3), it is possible to demonstrate that the purity is only locally affected since, once integrated, one has

$$\frac{D\mathcal{P}}{Dt} \propto \int_{-\infty}^{+\infty} d\xi_p \, W \left(\frac{\partial}{\partial \xi_p}\right)^{2k+1} W(\xi_x, \xi_p; t) = 0, \tag{D.12}$$

i.e. the purity is a constant of the motion if the integration volume is extended over all the phase-space, or even over a symmetric interval in the momentum direction, for the cases where W is symmetric in ξ_p.

In what concerns the quantum cosmological solutions discussed in Chapter 5, a quantifier of quantum fluctuations is given by the averaged value of $\Delta J_p(\xi_{x_c}(t), \xi_{p_c}(t); t)$ (corresponding to $\Delta J_p^\alpha(x, p; \tau)$ for the cosmological canonical variables) over all the phase-space volume. For periodic motions defined by classical trajectories parametrized by $\mathcal{C}$ [178], one can assign to the classical trajectory the role of a two-dimensional boundary contour for the Wigner flow. In this case, the integration of the dependent expression for $\partial W/\partial t$ over a volume $V_\mathcal{C}$ enclosed by $\mathcal{C}$, results into

$$\int_{V_\mathcal{C}} dV \frac{\partial W}{\partial t} = \int_{V_\mathcal{C}} dV \left(\frac{DW}{Dt} - \mathbf{v}_{\xi(\mathcal{C})} \cdot \boldsymbol{\nabla}_\xi W\right) = \frac{D}{Dt} \int_{V_\mathcal{C}} dV \, W - \int_c dV \, \boldsymbol{\nabla}_\xi \cdot (\mathbf{v}_{\xi(\mathcal{C})} W), \tag{D.13}$$

which, upon substitution into an integrated version of Eq. (D.1), leads to

$$\frac{D}{Dt} \int_{V_\mathcal{C}} dV \, W = \int_{V_\mathcal{C}} dV \left(\boldsymbol{\nabla}_\xi \cdot (\mathbf{v}_{\xi(\mathcal{C})} W) - \boldsymbol{\nabla}_\xi \cdot \mathbf{J}\right), \tag{D.14}$$

which vanishes in the classical limit, i.e. when $\mathbf{u} \sim \mathbf{v}_{\xi(\mathcal{C})}$.

To identify the quantum corrections in terms of $\Delta \mathbf{J} = \mathbf{J} - \mathbf{v}_{\xi(\mathcal{C})} W$, one computes the variation of the integrated probability flux enclosed by the classical surface, $\mathcal{C}$, in terms of an integral over a path given by

$$\frac{D}{Dt} \int_{V_\mathcal{C}} dV \, W = \frac{D}{Dt} \text{Prob}_{(\mathcal{C})} = - \int_{V_\mathcal{C}} dV \, \boldsymbol{\nabla}_\xi \cdot \Delta \mathbf{J} = - \oint_c d\ell \, \Delta \mathbf{J} \cdot \mathbf{n}, \tag{D.15}$$

where the unitary vector, $\mathbf{n} = (-\dot{\xi}_{pc}, \dot{\xi}_{x_c})|\mathbf{v}_{\xi(\mathcal{C})}|^{-1}$, is orthogonal to $\mathbf{v}_{\xi(\mathcal{C})}$. By following the parameterization of the line element, $d\ell \equiv |\mathbf{v}_{\xi(\mathcal{C})}| dt$, one has

$$\frac{D}{Dt} \text{Prob}_{(\mathcal{C})} \bigg|_{t=T} = - \oint_c d\ell \, \mathbf{J} \cdot \mathbf{n} = - \int_0^T dt \, \Delta J_p(\xi_{x_c}(t), \xi_{p_c}(t); t) \, \dot{\xi}_{x_c}(t), \tag{D.16}$$

where $\xi_{x_c}(t)$ and $\xi_{p_c}(t)$ are typical classical solutions, T is the period of the classical motion, and $\Delta J_p(\xi_x, \xi_p; t)$ is given by Eq. (D.3) for $k \geq 1$.

9 783384 251930